Volume 11

SCIENTIFIC PROGRESS

SCIENTIFIC PROGRESS

JAMES JEANS, WILLIAM BRAGG, E. V. APPLETON,
E. MELLANBY, J.B.S. HALDANE
AND JULIAN HUXLEY

LONDON AND NEW YORK

First published 1936 by Routledge

2 Park Square, Milton Park, Abingdon, Oxon OX14 4RN
711 Third Avenue, New York, NY 10017, USA

First issued in paperback 2016

Routledge is an imprint of the Taylor & Francis Group, an informa business

British Library Cataloguing in Publication Data
A catalogue record for this book is available from the British Library

ISBN-13: 978-1-138-01359-9 (hbk)
ISBN-13: 978-1-138-98146-1 (pbk)
ISBN: 978-0-415-73519-3 (Set)
eISBN: 978-1-315-77941-6 (Set)
eISBN: 978-1-315-77929-4 (Volume 11)

Disclaimer
The publisher has made every effort to trace copyright holders and would
welcome correspondence from those they have been unable to trace.

SIR HALLEY STEWART LECTURE, 1935

SCIENTIFIC PROGRESS

by

Sir James Jeans, F.R.S.

Sir William Bragg, O.M., F.R.S.

Professor E. V. Appleton, F.R.S.

Professor E. Mellanby, F.R.S.

Professor J. B. S. Haldane, F.R.S.

Professor Julian Huxley

Illustrated

LONDON

George Allen & Unwin Ltd

MUSEUM STREET

FIRST PUBLISHED IN 1936
CHEAP EDITION 1939

PRINTED IN GREAT BRITAIN BY
UNWIN BROTHERS LTD., WOKING

CONTENTS

I

SIR JAMES JEANS

F.R.S.

CHAPTER I

MAN AND THE UNIVERSE

A LARGE part of science is purely utilitarian; in the jargon of popular journalism, it is concerned with "harnessing the forces of Nature to the service of man." We find out what we can about the workings of Nature in order that our lives may be richer, more pleasurable, more comfortable, and more painless. But there are other branches of science which are purely intellectual; they try to understand, rather than to harness, the forces of Nature; their contribution is to gratify our intellectual curiosity as to the universe in which our lives are cast, and help us assess and understand our relation to that universe.

Until quite recently, these sciences, and indeed science as a whole, appeared to be continually hostile to man's good opinion of himself. Time after time new scientific knowledge compelled him to reconsider his beliefs as to his own position and importance in the universe, and each revision of his views proved very damaging to his self-esteem. But my main thesis to-day will be that, within the last few years, the tide has turned. We are now entitled to think rather better of our position in the universe than seemed possible one or two decades ago.

Let us briefly survey the outstanding landmarks in the history of this problem. We need not spend time over the speculations of remote antiquity; these were not scientific at all, and merely betrayed humanity into believing what it most wished to believe—into building for itself a city of dream-castles, so designed as to form a pleasant refuge from the injustices, disappointments, and perplexities of everyday life. Let us rather begin with the publication of a book—*De Revolutionibus Orbium Coelestium*—by a Polish monk, Copernicus, in 1543. In this book Copernicus argued that the sun does not move round the earth, but the earth round the sun. It was not a new thesis; Aristarchus of Samos had said the same thing 1,800 years earlier. His views had not, however, obtained any measure of general acceptance, so that at the time when Copernicus' book appeared most men thought of this little earth of ours as the centre of the whole universe, the sun and myriads of stars existing for no purpose but to circle round it, and dance attendance on it. The new theories of Copernicus opened the road to our present knowledge—that the earth is only a tiny speck of dust, circling with other specks of dust round a sun which, although a million times bigger than our earth, is yet only a grain of sand in the vastness of space. We know now that there are about as many stars in space as there are grains of sand on all the sea-beaches in the world; our home in the universe is only a minute fragment broken loose, like many others, from one such grain of sand.

The arguments of Copernicus might well have fallen on barren soil, like those of Aristarchus eighteen hundred years earlier, had not the telescopic discoveries of Galileo provided visual confirmation of their truth, or at least of their extreme plausibility.

Something else, too, came to their support —a better understanding of the workings of Nature. This originated in the first instance in the researches of Galileo, and his famous experiments performed on, or rather off, the leaning tower of Pisa, but as our landmark we may suitably choose the publication of a second book, Sir Isaac Newton's *Principia Mathematica*, in 1686. In this Newton explained his law of gravitation, and when once the significance of this law was fully understood, it became exceedingly difficult to deny the Copernican cosmology. Efforts were still made to explain it away as a mere conjectural hypothesis, an alternative to the pre-Copernican views which had placed the earth at the centre of the universe, but, in the light of the theory of gravitation, each new astronomical observation became a witness to the truth of the Copernican theory and to the untenability of all opposing theories.

Looked at in its broad implications, the law of gravitation was important, not so much because it told us why an apple fell to the ground, or why the earth and planets moved round the sun, as because it suggested that the whole of Nature was governed by hard and fast laws. For instance, to the ancients comets had been fearsome portents of evil, of famines or pestilences, of wars or the death of kings; seen in the light of Newton's work, they became mere inert chunks of matter, following their predestined paths as they were dragged about in space by the gravitational pull of the sun. Clearly, their motions could have nothing to do with the deeds or misdeeds of men. In the same way, darkness spread over the earth at an eclipse of the sun, not because the gods were angry with men, but because gravitation had pulled the moon into a position in which it temporarily shut off the sun's light—a position to which it had been predestined from the beginning of time. The tyranny of superstition and magic was broken, and Nature became something to study, not something to fear. Man began to see that he was free to work out his own destiny without fear of disturbance from interfering gods, spirits, or demons.

Yet it soon became clear that man could only claim to work out his own destiny by arrogating to himself precisely those powers that he was now denying to his dethroned gods—the power of interfering with the predestined course of Nature. If the gods could not push the comets where they wished, why should man be able to push a dagger or a dose of poison where he wished—or refrain from pushing them? If the moon was compelled to blot out the light of the sun, why should not men be compelled to push daggers and poisons about in the same way? Why hold murderers responsible for their actions if they were only fulfilling the predestined course of Nature? Out of such conjectures grew the determinist philosophy, which regarded human lives as mere irresponsible cogs in the vast machine of the universe. The philosophy itself was, of course, only a bundle of conjectures. Living matter was obviously in some way different from inanimate matter, and might well be subject to different laws.

We must step aside at this stage of our record to mention the publication of yet a third book, Charles Darwin's *Origin of Species*, in 1859. Man had so far thought of himself as a special creation, created *after* all other living things, and given dominion over them by divine right. Henceforth he was to see himself merely as the most successful member of one vast family which included—as his more lowly and less successful blood-relations —all kinds of animals, pleasant and unpleasant, fishes, reptiles, and creeping things. These last were the old-established denizens of the earth; they had inhabited it for hundreds of millions of years, but man—a mere parvenu—for less than one million. Man's success had come to him, so to speak, only yesterday; before then he had fought with the wild beasts and had not always prevailed. And he still has to fight for his position. If he fails in the fight, he may be exterminated, like the mammoths and dinosaurs before him, by some race of smaller beings—possibly by microbes too small for him even to see. If he succeeds, his capacities may gradually be extended and increased until he develops into something entirely different from, and perhaps entirely superior to, his present self. Our posterity of ten million years hence may well differ from us of to-day by as much as we differ from our lowly ancestry of ten million years ago. The worst that our race has to fear is defeat and extermination; the best it has to hope is to become a forgotten stepping-stone on the road to higher things.

So far, then, the progress of science had compelled man to rate his own importance continually lower and lower in the scale. He had started by fondly imagining that his home in space was the centre of the universe, and himself the central fact of the whole creation. By gradual steps he had come to see himself as a denizen of an inferior planet, which was something less than a speck of dust in the vast profusion of matter in the sky. Even on this inferior planet, he saw himself as only one in a long sequence of animals moving slowly—inconceivably slowly—towards perfection or disaster, possibly as a mere cog in the wheel, whose only function was to fulfil a destiny which had been determined before the commencement of time, and over which he had no control.

It is difficult to imagine human importance being rated lower than this, yet many thought that the physical theory of relativity, which Einstein advanced in 1905, exhibited human life in a still more ignoble light. Hitherto, the scientist and the plain man had been at one in thinking that events came to maturity with the passage of time, somewhat as the pattern of a tapestry is woven out of a loom. At one instant so much of the pattern has been woven, an instant later a little more, and so on. The pattern of the yet unwoven part of the picture may be inevitably determined by the way in which the loom is set, or it may not; at any rate this part of the picture is not yet in existence. And so long as the weaving is not yet an accomplished fact, it is at least conceivable that something may still happen to modify it. The operator who works the loom can still alter the setting of the loom, and so, within limits, modify those parts of the tapestry which are still to come, according to his choice. In the same way, it seemed possible that humanity, and life in general, might be able to exercise some influence, however slight, on events that had not yet emerged from the womb of time.

Then the theory of relativity came and taught that there is no clear-cut distinction between space and time; time is so interwoven with space that it is impossible to divide it up into past, present, and future in any absolute manner. This being so, the tapestry cannot consistently be divided into those parts which are already woven and those which are still to be woven. Such a distinction can have no objective reality behind it; at best it can have a meaning only for an individual personality, and then the distinction is between those parts he can see to be woven and those parts he cannot see.

The shortest cut to logical consistency was to suppose that the tapestry is already woven throughout its full extent, both in space and time, so that the whole picture exists, although we only become conscious of it bit by bit—like separate flies crawling over a tapestry. It is meaningless to speak of the parts which are yet to come—all we can speak of are the parts to which *we* are yet to come. And it is futile to speak of trying to alter these, because, although they may be yet to come for us, they may already have come for others. Such a view reduced living beings to automata. From being a creative machine, at any rate in its own estimation, the human consciousness declined to being a mere recording instrument. It could no longer claim any affinity with either the artist who designed the tapestry, or the craftsman who realized the design. A human life was reduced to a mere thread in the tapestry. As the flow of time dragged its consciousness over these threads, it might register horror, pity, or satisfaction at what it saw, but only as a spectator

at a cinema, who feels his emotions stirred by
what he sees on the screen. The pictures influence
him, but he cannot influence the pictures—he
has no more influence over the pictures yet to
come than a barometer has over the weather
yet to come.

I do not think it can be maintained—or even
has been very seriously maintained—that such
a view is absolutely forced upon us in any compelling manner by the facts of physics. At one
time it seemed plausible because it gave a simple
explanation of these facts, but no one would
maintain that it is a unique explanation. However
this may be, the whole situation is changed when
we pass from the phenomena of physics to those
of astronomy.

The purely physical theory of relativity recognizes no clear-cut distinction between space and
time, since it can explain all purely physical
phenomena without calling upon any such distinction. It cannot, however, explain the phenomenon of gravitation.

In 1915 Einstein tried to extend the theory of
relativity so that it should cover the facts of
astronomy and of gravitation in particular. The
simplest explanation of the phenomena seemed to
lie in supposing space to be curved, and the
universe as a consequence to be finite in its spatial
extension. It was natural to try in the first instance
to retain the symmetry between space and time
which had figured so prominently in the simpler

physical theory, but this was soon found to be impossible. If the theory of relativity was to be enlarged so as to cover the facts of astronomy, then the symmetry between space and time which had hitherto prevailed must be discarded. Thus time regained a real objective existence, although only on the astronomical scale, and with reference to astronomical phenomena.

De Sitter attacked the problem from the other end, and reached the same conclusion. He began by postulating symmetry between space and time, and found that for such symmetry to prevail the universe must be totally devoid of matter. As the actual universe certainly contains matter, de Sitter's postulate must clearly be discarded. In other words, space and time must be intrinsically different in their natures—which is, of course, precisely what the plain man has believed throughout.

To round off this part of our discussion, it must be added that the problem of the general nature of the universe has proved to be far more intricate than its early pioneers, Einstein and de Sitter, imagined. The two universes which they proposed are now known to be impossible, and have been discarded—de Sitter's because it could contain no matter and Einstein's because a universe of the type he imagined could have only a transitory existence; it would immediately start either expanding or contracting. Meanwhile, mathematical analysis has shown that an infinite number of other types of universe are mathematically possible. And these all have two properties in common.

In the first place, in every one of them space is either expanding or contracting. The observed recessions of the extra-galactic nebulae confirm this prediction of theory; these nebulae are galaxies of stars of the same general structure as our own, and all are observed to be receding from our own galaxy with speeds which are proportional to their distances from us—which is precisely the law which our mathematical theories lead us to expect.

This, at least, encourages us to think that our theory is on the right lines. Now, the second property which all the mathematical solutions have in common is that every one of them makes a real distinction between space and time. This gives us every justification for reverting to our old intuitional belief that past, present, and future have real objective meanings, and are not mere hallucinations of our individual minds—in brief we are free to believe that time is real.

Let me try to clarify the present situation by a simple, although I fear somewhat ludicrous, analogy. Imagine a race of worm-like beings, confined to living underground; they are endowed with senses which are sensitive to the purely physical forces of electricity, light, radiation, and so forth, but on which gravitation produces no effect. Now the laws of electricity, light, and radiation make no distinction between horizontal and vertical, so that, so far as the senses of these beings inform them, there are simply three symmetrical directions of space—just as in the real universe the laws of light, electricity, and radiation tell us only of four symmetrical dimensions, which we divide up subjectively into space and time. In spite of this, the worm-like beings might feel convinced that one direction, which they provisionally described as the vertical, was somehow essentially different from the remaining two, which they called horizontal. Neither through their senses, nor through the physical instruments which extend the range of their senses, can they detect any difference, but nevertheless they feel it must be there. We may appropriately bring this foolish little parable to an end by supposing that a small group of worms, more enterprizing than the rest, climb upwards until, like Dante climbing out of hell, they reach the upper air and see the stars. They at once see that there is a real distinction between horizontal and vertical; what physics failed to disclose, astronomy reveals. It is the same in the theory of relativity; we find a distinction between time and space, as soon as we abandon local physics and call the astronomy of the universe to our aid.

All this ought to reassure us that human lives are something more than mere threads in an already woven tapestry. We may still be cogs in a machine, but if so, the machine is in operation, it is moving, changing, and living—in so far as a machine can live.

And now comes the final chapter of our story, which carries us back again from astronomy to physics. Our last milestone was the formulation of the theory of relativity in 1915; our final landmark is the formulation of the new quantum-mechanics, and especially that branch of it known as wave-mechanics, in and about the year 1926. I am afraid it is not easy, without going into technicalities, to explain why this branch of science should exist at all.

Until the year 1926 scientists had regarded the universe as a collection of material objects which moved about in space, much as aeroplanes move about in the atmosphere. Experimental study suggested that all objects could be resolved in the last resort into particles of infinitesimal size, and as the majority were electrically charged, their motion would necessarily result in the emission of radiation. The emitted radiation was, in actual fact, found to be of a very special and distinctive kind. This presented a real and definite problem to the physicist—to discover how the fundamental particles of matter were arranged, and how they moved, so as to emit the radiation actually observed. Each of the various arrangements and motions which could be imagined would give its own special type of radiation—which of them would give the radiation actually observed?

Until the year 1913 everyone who thought

about this problem took it for granted that the
fundamental particles must move continuously
in space, under deterministic laws. As soon as
these assumptions were abandoned, but not
before, real progress began to be made. Tremendous advances were made in two investigations,
one by Bohr of Copenhagen in 1913, and one by
Einstein in 1917. In the first of these, Bohr
abandoned the idea of continuous motion, and
showed that some of the simpler spectra, that of
atomic hydrogen in particular, could very naturally be explained as the result of discontinuous
motion; he imagined that the electron in the
hydrogen atom jumped discontinuously from
point to point in space. Four years later Einstein
showed that the radiation from a solid body could
be explained on the same lines, although only
provided the idea of determinism was also thrown
overboard. He imagined that the jumps of the
electron did not conform to any mechanical or
deterministic laws, but merely to laws of a statistical nature. Such a hypothesis was not entirely
unacceptable, since Rutherford had already found
evidence of similar laws and motions in the
phenomena of radio-activity.

Yet even these ideas, revolutionary though
they seemed, failed to explain certain more
complicated kinds of radiations. To do this, it
was found necessary to introduce further new
ideas, many of which proved to be of a highly
artificial character. Many physicists felt that the

whole theory was becoming too artificial and too unreal to be satisfactory.

At this point de Broglie and Schrödinger introduced the new quantum mechanics in 1926. This broke away completely, not only from the new conceptions of Bohr and Einstein, but also from what were to all appearances the impregnable facts of experimental physics. The universe was no longer regarded as a collection of particles, but as a collection of waves—particles no longer appeared in the picture at all. The new picture was not a mere amplification of the old; it ignored the old altogether, wiping out at a single stroke all the individual particles which experimental physics had studied with so much care, and starting afresh on a new canvas. And this canvas was no longer the space and time within which earlier science had tried to confine the universe; the new wave-picture passed entirely beyond ordinary space, although not beyond ordinary time. The waves of the wave-picture can, it is true, be represented in a space of a great number of dimensions, but this is a purely mathematical space, and is something entirely different from the three-dimensional space of everyday life.

The supposition—so often tacitly introduced into science—that the universe is confined within the limits of space and time, is, of course, a pure assumption, for which no *a priori* justification can be found. Space and time are known to science as the vehicle through which signals from distant objects are transmitted, in the form of electromagnetic waves, until they reach our senses, bringing us our knowledge of the external world of Nature, but we have no right to assume that this external world is itself confined within the limits of space and time. The human mind is acquainted with the external world only through these signals, and so was tempted to assume, until quite recently, that the whole of reality was confined within the limits of the framework occupied by the signals. Such an assumption, as we now see, was a pure conjecture, and so can be judged only by the consistency of the conclusions to which it leads, and the fidelity with which these can reproduce the phenomena of Nature. Bohr's theory of atomic spectra, to which I have already referred, was in effect an attempt to depict the atom as a structure confined within the limits of space and time. But it led to a difficulty, and almost an inconsistency, at once, since it was based upon the concept of discontinuous motion in space and time. The idea of an electron suddenly moving from one point of space to another with infinite velocity is repellent both to the intuition of the plain man and to the fundamental knowledge of the physicist, so that we are driven to think of the electron as the manifestation in space of something which exists outside space—somewhat as we might see the head of a swimmer disappear, while his feet simultaneously rose above the surface of the water

at a different place. We should not conclude
that something had moved over the surface of
the water at infinite speed, but rather that something existed below the surface of the water.
This simple analogy leads us to reflect that space-
time may be only a sort of outer surface of reality;
although the waves which bring us our knowledge
of reality travel as ripples on the surface, yet they
seem themselves to bring us messages of depths
beneath. Thus it is no reproach to the wave-
picture that it does not attempt to depict the
universe on a canvas of time and space.

Although the wave-picture has been surpris-
ingly successful in explaining a variety of physical
phenomena, it has not banished the particle-
picture from science; it would be more true to
say that since the wave-mechanics first appeared
in 1926, there have been two distinct pictures
of nature in the field, one trying to depict nature
as waves, and the other as particles.

Neither picture is entirely complete, and it has
gradually become clear that the material world
consists of something which cannot adequately
be described as either waves or particles. It is
obviously something which cannot be grasped in
its entirety by the human imagination, so that
the best we can do is to represent it by pictures,
each of which contains a partial, but only a
partial, approximation to the whole truth. We shall
realize how far we are from a full understanding
of this something if we pause to reflect how

different a particle is, in all its properties and qualities, from a wave. These are, so to speak, projections of the reality on different planes of thought, and we shall only understand the full reality when, if ever, we can see one gradually transforming into the other. As I write, the sun is shining, and I hold in my hand an ordinary pair of spectacles. Holding it in one way its shadow on my paper consists of the three sides of a square ⊔; when I turn it through a right angle its shadow consists of two circles ᴏᴏ. But a being whose thoughts were limited to two dimensions would be unable to imagine the object being turned through a right angle, or to watch the gradual metamorphosis of one shadow into another; such a being might find it hard to realize that the two very different shadow-pictures were projections of the same object. In the same way we must move to some new plane of thought before we can realize that the particles and the waves are shadow-pictures of one and the same universe.

Happily, we know to what new plane of thought we must move. It is the mathematical. The mathematician has a specification of the constituent parts of the universe which he believes to be fairly complete. Look at it with care, and you will see that the universe consists of particles. Now let the mathematician go through the process which he will describe technically as "changing the variables," and look again. Although it is still the same specification of precisely the same universe, it is presented in a much altered form, and we see now that it represents a series of waves—the particles have disappeared from the picture as completely as the straight lines disappeared from the shadow when I turned my spectacles through a right angle.

It follows from all this that if we want the ultimate truth about the universe or its constituents, we must go to the mathematician. The experimental physicist and the astronomer can each, no doubt, tell us a great deal, but only within limits. We ask the experimental physicist to tell us by how much the more massive of the two particles which constitute the hydrogen atom is heavier than its companion electron. He works in his laboratory for a long time with spectroscopes, gratings, and so forth, and emerges to tell us it is somewhere between 1847 and 1848. We ask Sir Arthur Eddington, and, without going into a laboratory at all, he tells us it is the ratio of the roots of a simple quadratic equation:

$$x^2 - 136x + 10 = 0$$

It begins 1847·6 . . . , and if we will give him a few minutes he will calculate it to as many places of decimals as we like. The point I want to bring out is that the methods of the mathematician can give us a full and final answer, while those of the experimentalist only give a partial answer.

Again, we ask the practical astronomer how many particles there are in the universe. He will try to estimate the weight of one of the great extra-galactic nebulae from the pull it exerts on the stars of which it is composed. He will multiply this by the number of nebulae he can see in his telescope, add something at a guess for the nebulae he cannot see, and give us a very big number for his answer—which may well be a thousand times too large or a thousand times too small. We ask Sir Arthur Eddington, and, without going to a telescope at all, he will tell us that he thinks the number is very possibly 136×2^{256}, not a single particle more or less. We are not concerned with this or any other particular number—it would make no difference to the argument if Eddington should announce tomorrow that when he had said 136 he meant 137. The essential point is, these fundamental questions are the province of the mathematician; they can be answered only by mathematical means. In brief, we live in a mathematical universe.

This being so, let us consult the mathematician about some of the fundamental problems we discussed before, particularly about the problem of whether or not events in the material universe are strictly determinate, so that one follows inevitably from another and the whole course of the universe is implied in the initial arrangements of its constituents. To this question, the particle-picture and the wave-picture seem at first to suggest very different answers; it is, of course, the business of the mathematician to reconcile their apparent divergencies.

When we regard the material world as made up of particles, we have to suppose that the motion of these particles is not fully determinate—otherwise we cannot obtain agreement with observation. The motion of any one single particle does not appear to obey any definite determinate laws, although the motion of a great number of particles is found to obey quite definite statistical laws. Indeed, it is from such statistical laws that the apparent uniformity of Nature proceeds. If, then, we confine ourselves to the particle-picture, we inevitably conclude that Nature in the last resort is not strictly determinate—somewhere there is an opening for new features to appear in the universe, although we do not know from whence they proceed or what causes them.

This does not, of course, prove that nature is not deterministic, because we have no right to confine ourselves to the particle-picture, thereby assuming that Nature consists of particles. Nevertheless, it does provide a rebuttal to the old argument of the materialists. Their argument was, in brief, as follows:

"Nature is completely determinate, because it consists of particles, each of which moves about in space under unbreakable laws such that the motion of each particle is determined solely by the previous state of the system."

To which we can now reply:

"It is true that your picture of Nature as a collection of particles is a good picture in that it reproduces many phenomena with great fidelity, but if you picture Nature as particles, modern physics shows that their motion is not deterministic, but the reverse. Thus, on your own premises, you are wrong."

Let us now turn to the wave-picture, which seems at first sight to tell us that Nature is strictly deterministic. In this picture the waves spread like ripples on the surface of water. Now such ripples follow a predetermined course, and from a knowledge of the ripples at any one instant, the mathematician can calculate what the motion of these ripples will be, and so what the configuration of the surface will be at any future instant. The wave-picture of Nature has precisely the same properties. Knowing the waves at any instant, they can be calculated at any future instant; in other words, they are completely determinate.

It may seem paradoxical that the two pictures of Nature should seem to convey such different messages, but any paradox there may be disappears when we probe deeper into the physical meaning of the waves. We then find that the two pictures are trying to depict entirely different things. The particle-picture assumes that an objective universe exists outside ourselves, and tries to depict this objective universe, while the wave-picture tries to depict our knowledge of the universe as experienced by us.

We spoke just now of the mathematician "changing his variables" as he passed from the particle-picture to the wave-picture. His first set of variables were the positions of the particles which, according to the particle-picture, constitute the objective universe. His second set of variables was our knowledge of these supposed particles, as derived from observation. In brief, the wave-picture is a picture of knowledge, not of things. The textbooks tell us that the wave-mechanics depict an electron as a system of waves, but when we ask what particular waves represent any particular objective electron, no answer is forthcoming; we must first say precisely how much we know about the electron in question. Then we find that the waves do not depict the electron, but what we know about the electron. They depict something inside our minds, not something outside. Thus the wave-mechanics gives no confirmation to the supposition that the electron has an objective existence independently of our knowledge of it—rather it suggests that our knowledge of the universe (or the bit of it which we experience) is fundamental, and that the electron is a clumsy creation of our own, resulting from our efforts to locate our universe in space and time. This being so, we can see why the wave-picture gives the best representation of ultimate reality, as experienced by us; we also begin to suspect that the particle-picture is only an artificial model, which is useful as a concession to the limitations of the human mind.

In the light of all this, we see that the apparent indeterminism of the particle-picture becomes meaningless, while the determinism of the wave-picture has nothing to do with the course of objective Nature. The determinism is not one of motions or configurations of parts of the universe, but merely of our knowledge of these motions and configurations. It is obvious that by acquiring new knowledge of the universe—as, for instance, by performing an experiment on any small part of it—we can change the total of our knowledge, and so change the waves which represent this small part of the universe in the wave-picture. The determinism of the wave-picture does not proclaim that the course of Nature follows undeviating and unchangeable laws, but merely that our knowledge of Nature can only be altered by the acquisition of new knowledge. Thus it tells us nothing at all as to the actual course of Nature, and we have to conclude that the science of to-day is unable to give any decision—is unable even to produce any evidence—in the long-debated question of determinacy. We still may or may not be automata; present-day physics certainly cannot prove that we are not, but it gives us no shadow of a reason for thinking that we are.

Reviewing the history of the problem as a

whole, we have seen that until early in the present century scientific knowledge was continually compelling man to lower his estimate of his importance and of his position in the universe. It is still too early to attempt any final judgment of more recent events, partly because they are still so near at hand, partly because scientists themselves are not yet in complete agreement about them. But to me, at least, it seems that within the last few years the tide has begun to turn. In the light of recent knowledge gained from the theory of relativity and quanta we seem entitled to take a more hopeful view of our position than Victorian science had been willing to concede. The plain average man, ignorant alike of the problems and of the perplexities of science, had formed a view of his position which was influenced largely, no doubt, by his vanity and self-importance, but was based, on the whole, on his practical everyday experience of life. He believed, among other things, that he was free to choose between the higher and the lower, between good and evil, between progress and decadence. To many, Victorian science seemed to challenge all such beliefs. It knew nothing of higher nor lower, progress nor decadence; it knew only of a vast machine, which ran on automatically and of its own inertia, as it had been set to run on the first morning of the creation. And it would continue so to run, following out its predestined course, to the end of time.

We now begin to think that this challenge was a mistaken one, that the universe may be more like the untutored man's commonsense conception of it than had seemed possible a generation ago, and that humanity may not have been mistaken in thinking itself free to choose between good and evil, to decide its direction of development, and within limits to carve out its own future.

II

SIR WILLIAM BRAGG

O.M., F.R.S.

CHAPTER II

THE PROGRESS OF PHYSICAL SCIENCE

IF we give to the term Progress in Science the meaning which is most simple and direct, we shall suppose it to refer to the growth of our knowledge of the world in which we live. The first purpose of scientific enquiry is to add to the extent and the accuracy of that knowledge. Our methods are based on observation: we are content to watch what Nature does, to examine the materials which she uses, and to trace relations between her various operations. Our success depends upon the accuracy with which we can observe, and upon the number of comparisons which we can make. To many of us it is a continual pleasure to make these enquiries. There is the incentive that they are often difficult and baffling, sometimes yielding rich and unexpected results, more often giving no more enjoyment than that of the unsuccessful chase. There will always be found those who ask for nothing more than freedom to investigate and to learn. When I am asked to speak of Progress in Science, I cannot but think in the first place of the activities of the disinterested searcher to whom a growing knowledge of Nature is sufficient reward.

It has been suggested to me that I should also speak of the applications of scientific knowledge, and here we enter a very different field. Because we are ourselves a part of Nature's work and are caught up in her activities every increase in knowledge may give us an opportunity of bending those activities to our desires. As we know, there are many who are quick to take advantage of the fact: and their work is sometimes well done, sometimes ill. When we come to consider the applications of science we find ourselves in touch with problems of economies, of ethics, and, in general, of the behaviour of men to one another. In these later days the applications of science have become more obvious, and because of their great width and their power they interest an ever-widening circle in our community.

Let me first speak of the purely intellectual side of our subject, that is to say, of the progress of science itself. I am especially to relate what I have to say to the physical sciences. I need not make any apology to my chemical friends if I assume that I may deal with their work also, because our two sciences, that once could be kept apart more or less, are now increasingly blended into one.

We are all aware that the advancement of knowledge in the physical sciences has been exceedingly rapid in recent years. If we needed definite proofs of that statement we should only have to look around us and observe its consequences. Now we cannot in one hour, which I hope will seem to you to be reasonably short, consider the details of the advance: we can but try to form some idea of its trend. We may ask ourselves why the onward movement should so suddenly have increased its pace, and gone so far in the last few years.

No doubt there have been influences external to science itself which have been remarkably active. In particular, there has been a growing appreciation of the power that a knowledge of Nature puts into the hands of men; especially was this fostered by the experiences of the War. But the internal influences are of the greater interest from our present point of view.

When we look back on the progress of the physical and chemical sciences in recent years we observe that a few great generalizations have been of extraordinary importance. One of the greatest of these may be described in the statement that Nature is, in one of her chief aspects, essentially "particulate." Discontinuity is her usual wear, with an urge towards order and regularity.

Long ago the old philosophers guessed at something of the sort. Democritus and Lucretius supposed that if matter were sufficiently subdivided a stage would be reached when further division would be impossible: even if it could be accomplished the substance would be no longer what it was. An opposing theory looked on matter as capable of infinite subdivision without showing any change in properties. There is a tremendous difference between the two views. According to the former it is to be expected that enquiry into the minute structure of matter must be rewarded by a better understanding of natural processes: and this has indeed been the case. According to the latter, however minutely we examined, we should never learn anything new. Every substance in the world would be *sui generis*; there would be no common foundation or element of structure; scientific research would be barren indeed. Our minds would lack the opportunities of moving in rich fields of observation and thought whereby we might better ourselves and our conditions.

The atomic theory was in its early form dim and imperfect: it supposed that every substance was formed of atoms, but it missed the fundamental point that the number of different kinds of atom is very limited, and that the atoms of any one kind, as, for example, of gold or oxygen or silicon, are all exactly alike. There are reservations to be made with respect to that similarity, but for the moment they are of no consequence. This most important advance we associate mainly with the name of John Dalton. The advances in chemistry which were made in the nineteenth century were all based upon the newer and better understanding. The particulate character of matter was appreciated. It came to be realized that though the number of different kinds of atoms was small, and the number of kinds in common use was much smaller still, yet Nature achieved her rich variety by fitting the atoms together in an infinite number of ways, thus producing materials of widely varying qualities. The first constructive step is in general the formation of the molecule. A molecule is a small company of atoms resisting disintegration more or less, and therefore more or less permanent. In very many cases the number of atoms in the molecule is quite small. Thus the molecule of water is, as is generally known, a fairly permanent combination of two atoms of hydrogen with one of oxygen. Sometimes the number of atoms in the molecule is very much larger; in the case of the complicated protein molecule it may run to thousands. The atoms may be compared to the letters of an alphabet, each letter being incapable of division without loss of significance. Molecules are to be compared with words formed of letters: a very limited number of letters is sufficient for the formation of a very large number of words.

The assemblage of molecules to form solids and liquids is a further step in the construction of the materials of the world. The analogy must not be pressed to cover this also, because the molecules in such assemblages are often of one kind throughout.

It is well to form some idea of the magnitude of atoms and molecules, and this may be done by considering the arrangement of objects in the adjoining table, Fig. 1.

This shows the relative magnitudes of various objects which we observe and measure. It is like a set of shelves on which we place specimens of objects and magnitudes from the very great to

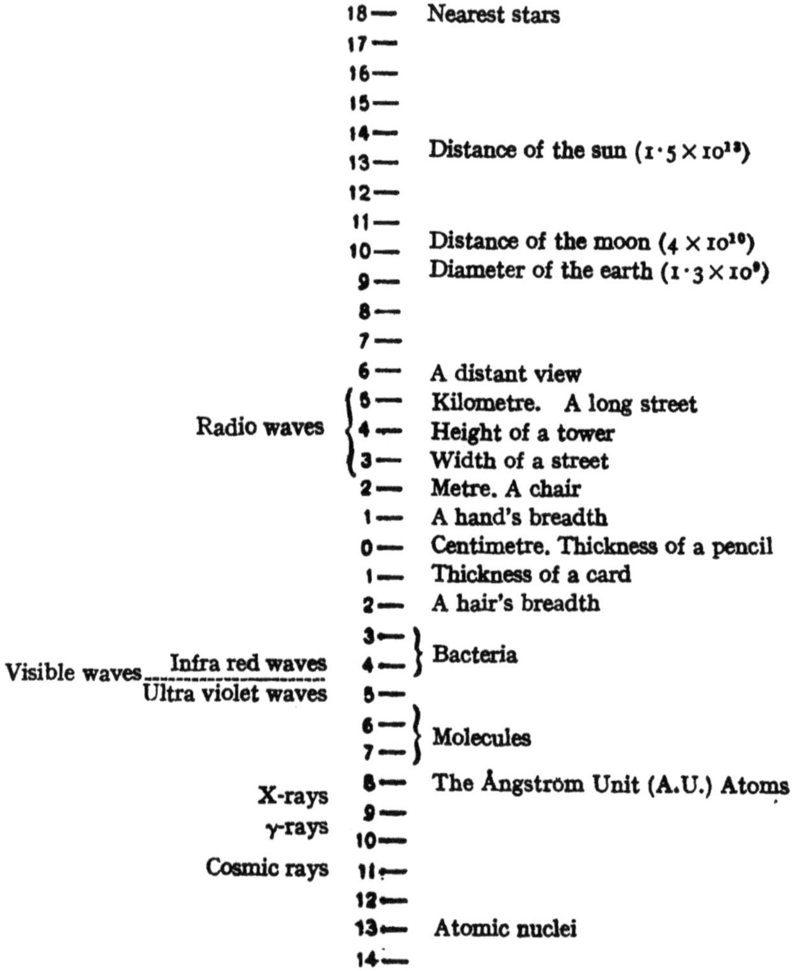

18 —	Nearest stars
17 —	
16 —	
15 —	
14 —	
13 —	Distance of the sun ($1 \cdot 5 \times 10^{13}$)
12 —	
11 —	
10 —	Distance of the moon (4×10^{10})
9 —	Diameter of the earth ($1 \cdot 3 \times 10^{9}$)
8 —	
7 —	
6 —	A distant view
5 —	Kilometre. A long street
4 —	Height of a tower
3 —	Width of a street
2 —	Metre. A chair
1 —	A hand's breadth
0 —	Centimetre. Thickness of a pencil
1 —	Thickness of a card
2 —	A hair's breadth
3 —	} Bacteria
4 —	
5 —	
6 —	} Molecules
7 —	
8 —	The Ångström Unit (A.U.) Atoms
9 —	
10 —	
11 —	
12 —	
13 —	Atomic nuclei
14 —	

Radio waves (bracketing 5—4—3)

Visible waves — Infra red waves / Ultra violet waves

X-rays, γ-rays, Cosmic rays

FIG. 1.

the very small. On a middle shelf marked zero we have the centimetre, and the thickness of a pencil to represent objects of that order of magnitude. On the shelf above we place an object of about ten centimetres in size; the width of a hand will serve. The shelf above takes objects of about a hundred centimetres, for example, smaller objects of furniture. The width of a street will represent the thousand centimetres, the height of a tower might be ten thousand centimetres or a hundred metres, and so on. Below the zero shelf comes first a shelf holding something of the order of a millimetre in thickness, as a card; then the hair's breadth on the next shelf, and so on. Bacteria are at various heights on the third and fourth shelves down; molecules on the sixth and seventh, atoms nearly down to the eighth. On the other side of the vertical line the various wave-lengths are shown in the same way. Distances are sometimes given in figures. The sun's distance is fifteen million million centimetres, or in symbols $1 \cdot 5 \times 10^{13}$. This goes, therefore, on the thirteenth shelf up.

The processes of Nature, as well in things that have life as in those that have none, depend continuously upon the formation of molecules and combinations of molecules, upon their dispersal, and upon their reassembly in new forms. Just so the compositor breaks up his type and gathers it again into words and sentences of new meaning. When the chemist discovered this plan his science grew rapidly in power and extent. He found himself able to put atom and atom together, to build up molecules possessing desirable properties and again to take them to pieces; he learnt how to encourage some processes

and to check others. Chemistry, developing on
these lines, has become one of the most powerful
of the tools in man's possessions. All this has
followed directly from the discovery of one of the
great ways of Nature. The chemist's usefulness
and power rests upon his act of recognition, and
his acceptance of the guidance which it has given
him. He has thought in Nature's way, and he has
his reward. As the Greek philosopher said, "The
supreme virtue is thought, and wisdom consists
in saying what is true, and acting according to
Nature, listening to her."[1]

Towards the end of the nineteenth century
came a new realization of Nature's preference
for the particulate. Electricity had hitherto been
thought of in the way that in ancient times some
had thought of matter, as capable of infinite
subdivision. The discovery of the electron
changed our conception of electricity precisely
as the discovery of the true character of the atom
had improved our conceptions of the nature of
matter. Not only were material substances com-
pounded of separate atoms, but electricity also
was composed of separate units, and the simplifi-
cation was here even more striking because there
was only one size of unit. Surely it is an arresting
fact that all the movements of electricity in the
world can be described in numerical terms with
no fractions. And again, when the fundamental

---

[1] Heracleitus. See Livingstone's *Greek Ideals and Modern Life*, p. 78.

principle is grasped the progress of science becomes rapid. The very great development of electrical theory during the last thirty or forty years is due in the main to the discovery of the electron. Once more we have learnt to think in Nature's way.

Having found that both matter and electricity are particulate, we are perhaps prepared for the further discovery that even energy can, in a certain sense, be described by the same term. There was some truth, after all, in Newton's corpuscular theory of light: the advocates of the wave theory had dismissed it too hastily. We find ourselves driven to think of the energy which the sun sends to us in the form of light and similar radiations as contained in separate parcels, which can be handed to the electrons and atoms struck by the radiation. We speak of this new development as a feature of the quantum theory. As is well known, the physical and chemical sciences have already derived great profit from it. It has important associations with the theory of relativity. We cannot let the wave theory go, though the quantum theory seems to our imperfect sight to be in contradiction to it: we must be content to accept both theories for what they are worth, and wait until we can see more clearly. Wave-mechanics, the new mathematics which Sir James Jeans mentioned, aims at putting the two in double harness. Meanwhile, we observe that again progress has become rapid because we have learnt to think on the right lines.

Matter, electricity, energy, each in turn has been found to be fashioned in the same strange way. Now on these three and their mutual relations the construction and the processes of the world may be said, in broad terms, to depend. There is therefore an importance of the first rank in the universality of the particulate.

We have the conception of a world built of separate parts, many in number but few in kind, as a bridge, for example, is constructed by the engineer who rivets together his various structural elements. This does not imply that each such element is so simple that its own parts cannot be analysed and studied. An atom is surely a complicated thing. But the complications of any one kind of atom, copper for example, are characteristic and sufficiently constant, so that the copper atom may be treated as a standard unit in building operations.

Roughly, one might say that the atoms correspond to the members of the engineer's structure and the electrons to the rivets which hold the members together. With a little greater licence one might compare the energy-quanta to the exertions of the workmen who insert or remove the rivets. In these terms the construction or alteration of the engineer's design is a fairly close analogue to physical and chemical operations, such as the melting of wax before the fire or the growth of a plant in the rays of the sun. Bonds are broken and reformed, and energy-quanta are derived from the radiant light and heat.

When we apprehend this great constructive scheme we are irresistibly drawn to the analysis of its details. We try to repeat the same processes of building, destroying and rebuilding; satisfying thereby our instincts of enquiry, and hoping to direct things to our profit. The progress of science depends on such efforts. The rapidity of recent progress has been due to our better apprehension: we have a clearer understanding of our position, and of the direction in which we want to go.

We may now go on to consider how we have contrived to be aware of these numerical characteristics of Nature, to measure the properties of an atom or an electron, to determine the structure of a molecule, and the manner in which molecules are packed together in a solid substance. Such matters are outside the range of our unaided senses. The difficulty lies in the minuteness of that which is to be observed. To use an old illustration, if a drop of water were magnified to the size of the earth each molecule would not be much larger than a football. Yet I may show you presently pictures some of which represent the action of a single atom, a single electron, even a single quantum, while others show accurately the form and dimensions of single molecules. These are recent achievements. I must first say

a few words about the older well-tried methods
to which we owe so much. It is, of course, to be
noted that minuteness of scale does not imply
ineffectiveness, nor warrant neglect: any more
than that the smallness of the pattern of a
weaving is of no importance because the piece
of cloth is relatively large. The characteristics of
the cloth depend entirely on the characteristics of
the unit of pattern, however small it is: and just
so the processes of the world depend entirely on
the characteristics of those minute units of
structure which we must in some way be able to
examine in detail if we wish to understand how
the world moves; still more if we want to control
the motion.

Of course, the progress of physical and
chemical science is intellectual, depending on skill
and judgment in the arrangement of observations
and in deducing relations between them. But the
observations must come first, and observations
require instruments. Robert Hooke was one of
the first to venture boldly into the field of enquiry
into the minute: his *Micrographia* (1665) contains
his extraordinarily interesting account of what
the newly invented microscope had shown him.
We all know that marvellous improvements in
optical construction have added enormously to
the capacity of the microscope, and that it is of
eminent importance to all research in science,
and indeed to all attempts to apply science to
various purposes. It is a very long way, however,

FIG. 2.—Hooke's microscope.

FIG. 3.—Hooke's drawing of a razor blade as seen by
the aid of his microscope.

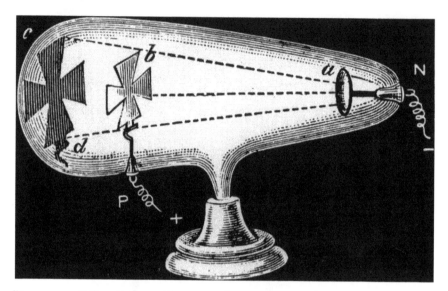

FIG. 4.—The figure shows a famous experiment devised by Crookes. In a tube exhausted almost completely of air, an electric discharge, in the form of a stream of electrons, is projected from the negative terminal *a*, and excites fluorescence in the glass of the opposite wall. A metal cross, *b*, casts a sharp shadow.

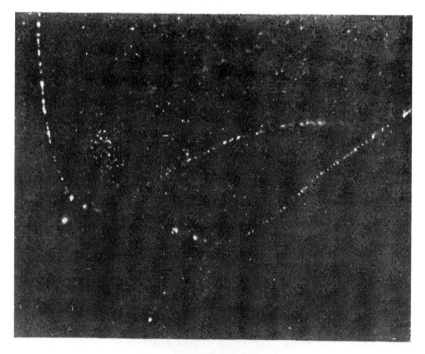

FIG. 5.—This figure shows fog tracks of electrons shot away from atoms of radium (C. T. R. Wilson).

from providing means by which a single atom can be seen, or its behaviour followed. The smallest object that can be examined in any detail contains millions of atoms. Nor is there any possibility of an improvement in the construction of the microscope which would take us nearer to those units we wish to see.

This has not, however, blocked the way entirely. The great expansion of chemistry which dates from the beginning of the nineteenth century has been due to the use of indirect means. The principal object of chemical research is to determine the rules which govern the association of atoms into molecules, and, in reverse, the dissociation of molecules into atoms. The investigation naturally includes the examination of the properties of the molecules when formed. If individual atoms are beyond the limits of vision by the microscope, how has it been possible to study the details of molecular construction? Help in the past came mainly from the observation that the elementary substances combined in definite proportions. When water is formed by the association of hydrogen and oxygen, the ratio of the weights of the two elementary substances must be exactly the same in all cases, if there are to be no leavings of either of them. Observed in thousands of instances without exception, this rule is simply explained on the supposition that the disappearances of two gases and the formation of a liquid is due to the formation of water molecules, each of which contains a definite number of each kind of atom. Further observation and ingenious reasoning justified the assumption that the water molecule contains two atoms of hydrogen to one of oxygen. On such arguments, continually tested and always satisfying, the present structure of chemistry has been built.

But throughout such observing and arguing it was difficult to find any clear determination of the weight or form or size or constitution of the single atom. Suddenly, there appeared at the end of last century three novel and powerful aids to further research: the electron, X-rays, and radio-activity. It then became possible to study not only the single electron, but also the single atom, the single molecule, and even the single X-ray, which is an energy-quantum. The electrons were discovered in the electric spark, which was simply a manifestation of the light and heat generated by their passage across an open space. The molecules of air or other gas are shattered and shaken by the swiftly moving stream. The movement must be swift, the electrons must be projected with tremendous speed, many thousands of miles a second. This can be achieved easily enough by well-known electrical means, and is seen on its grandest scale in the lightning flash.

Here were the electrons, but how were they to be grasped for the purpose of examination? The traffic was so dense and the number of collisions so great that the individuals were lost in the crowd. The number of collisions could be reduced if the number of molecules of the gas was reduced. Only when the design and details of the air-pump were sufficiently considered did success become possible. At a pressure of about a millionth of the normal the collisions are so much fewer that the passage of the electrons takes the form of a definite stream with recognizable boundaries, a stream like a jet from a hose-pipe (see Fig. 4, opposite p. 53). As soon as technique had reached this stage it became possible to examine the electron's properties. This was the accomplishment of the closing decades of the last century.

About thirty years ago a new and ingenious method of enquiry was devised, and it is this which I can use to show the effects of the single electron, atom, or X-ray. The swiftly moving particle may shatter atoms which it strikes, not injuring the nucleus, but removing electrons from the crowd that surrounds the nucleus. The path of the particle is strewn with these damaged atoms. Light and heat are produced, but the effect is far too small to be observed. Only when enormous numbers of electrons act together, as in the electric spark, is there anything to be seen. Now comes the ingenious part of a wonderful experiment. Advantage is taken of the fact that if there is water vapour in a space which is suddenly and sufficiently lowered in temperature, the vapour is deposited as dew or fog, as we are well accustomed to see. And the fog settles more readily on the "damaged" atoms than on anything else, because they are electrified and attract the water molecules.

If therefore an electron drives its way through air laden with vapour, and the air is chilled before the debris have time to disperse, a fog settles on the line of the track: with suitable arrangements the white line can be photographed.

The chill is effected very simply by allowing the air to expand suddenly, the containing chamber being enlarged to a suitable extent.

Examples of electron tracks are shown in Figs. 5 (opposite p. 53) and 8 (opposite p. 56).

We can hurl electrons about with speeds that make their motions observable. If we seek to treat atoms in the same way we have a far more difficult task. But we are relieved of that task by the fact that Nature is constantly performing it for us. When the radium or other radioactive atom explodes, the smaller part of it, actually an atom of helium, is blown away with a speed of about ten thousand miles a second, which is roughly a thousand times greater than the average speed of motion of the atoms of a gas among themselves, under normal conditions. Endowed with such a speed, the movement of the atom can be followed even more easily than that of the electron, because it damages so many more atoms on its way. It is also interesting to observe that the electron path is very devious because it is

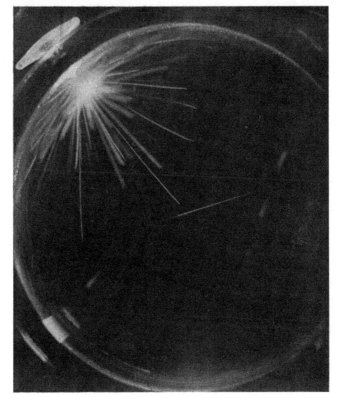

FIG. 6.—A minute quantity of radium is projecting helium atoms through the air in the fog chamber. The tracks are almost entirely straight lines. The diameter of the circular chamber is about 4 inches (C. T. R. Wilson).

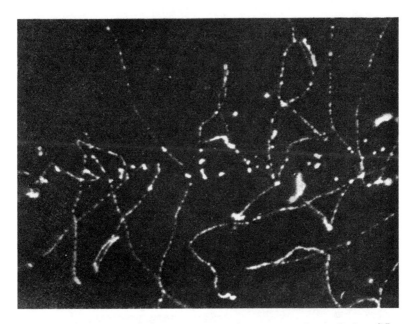

FIG. 8.—Fog tracks due to the electrons ejected by X-rays from the atoms of the gas in the fog chamber. The X-rays are contained in a fine stream moving across the chamber from right to left, but themselves leave no fog-trace (C. T. R. Wilson).

Fig. 10.—Forms of snow crystals: considerably
magnified.

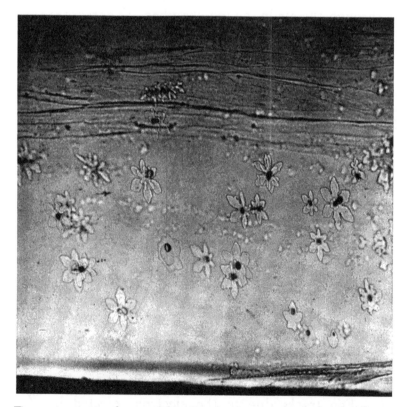

Fig. 11.—A piece of ice beginning to melt in the heat
of an arc lantern, the light of which projects the
picture of the ice-flowers upon the screen.

continually being turned aside by collisions with the nuclei of atoms. The helium atoms in the illustrations (Figs. 6 and 7) show no such deviation except rarely and at the end of their paths. The atom is so much heavier than the electron that it drives straight through the atoms which it meets.

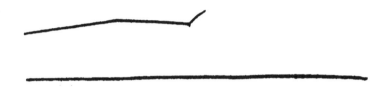

FIG. 7.—Enlarged pictures of two fog-tracks of helium atoms projected by radium (C. T. R. Wilson).

Thus we may now see the individual electron and the individual atom, not, of course, the entities themselves, but their immediate effects. Just so we say that we see a shooting star.

But we cannot observe the energy quantum of radiation in quite the same way. As long as it retains its form it produces no effect upon our eyes, or upon a photographic plate. It is only when the quantum strikes an atom that it disturbs or even ejects one of the electrons attached to the atom. Then something happens which can be registered. In our eyes the electron disturbance causes an effect which is transmitted to the brain. On the photographic plate there is a chemical effect. In the fog chamber the ejected electron, if swift enough, may leave a fog trail behind it. Consider, this example (Fig. 8 opposite p. 56) of the effects due to X-rays. The energy-quantum of the X-ray is of very much greater intensity than that of light, and when it strikes an atom it can eject an electron of sufficient speed. The picture shows many electron trails all starting from a narrow band which crosses the photograph, and includes the paths of the X-ray quanta that have gone by. Each electron trail begins at a point where an encounter between an atom and a quantum has taken place.

Now we come to the molecule. In this case also the individual may be shown, and its form and dimensions may be measured with accuracy; but the method to be employed differs entirely from those which we have just been considering. Let us remember that each molecule is an assemblage of atoms according to a definite pattern; and the properties of the molecule depend on both composition and arrangement. If we had sufficient knowledge of the properties of each atom we would be able to predict of any particular compound of atoms whether it would form a molecule, what shape the molecule would take, how the atoms would be arranged, what space it would occupy, what power it would have of resisting disintegration, what effect it would have on neighbouring atoms, and so on. Of course, we are very far away from the exercise of any such complete powers. The chemist has perforce been content to examine each kind of molecule individually. By a marvellous chain of reasoning he has been able to plan the atoms in the molecule, showing at the same time how the arrangement governs the molecular properties. In doing so he has built up the immense structure of modern

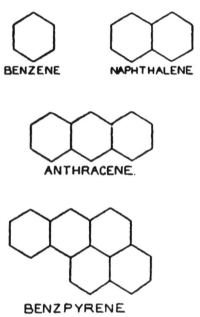

BENZENE NAPHTHALENE

ANTHRACENE.

BENZPYRENE

FIG. 9.—Skeleton diagrams of some organic molecules of great interest. In the benzene molecule six carbon atoms are arranged in a regular hexagon. To each carbon atom a hydrogen atom is attached, but is not shown. The length of the side of the hexagon is known very exactly. It is rather more than half of the hundred-millionth of an inch. Naphthalene and anthracene are well known in the dye industry. Benzpyrene is very important in the study of the causes of malignant disease. As in the case of benzene, the skeleton is formed of carbon atoms in these molecules also.

chemistry with its relations to all the other sciences, natural, biological, industrial. As I have already said, the arguments necessarily proceed by indirect ways because the individual atoms and molecules have, until lately, been beyond the powers of observation. The arrangement of the atoms in the molecule is not so well determined by the older methods as we could wish. It has, for example, been shown by chemical methods that the very important molecule of benzene consists of a ring of six carbon atoms, one hydrogen atom being attached to each carbon atom. But these methods do not give the dimensions of the ring, nor indeed do they tell us whether all the atoms lie in a plane. Nor do they tell us how the molecules are arranged when the benzene is frozen and solid, and that is a serious matter, because the relations between the behaviour of solid bodies and their structure interests us very greatly.

Let us remember that in all those constructions and destructions, arrangements and re-arrangements which constitute our changing world, two opposing principles are always at war. There is a constant tendency to aggregation because molecules draw together under the action of forces which are of an electrical character. There is a constant tendency to disruption because the continuous movements and vibrations due to heat would keep the molecules from settling down into definite places or positions. Thus, for example, the molecules of water, though always trying to associate, and indeed often succeeding temporarily, are prevented from forming any permanent structure because their heat keeps them on the move. If the heat is sufficiently reduced the water freezes.

Now this freezing is a most remarkable phenomenon. If water molecules merely settled down together like mud at the bottom of a pond there would be little of interest in it. But the actual fact is one of great beauty. The molecules link themselves together in a simple pattern of extreme regularity: the unit of pattern contains very few atoms, and is repeated exactly through the whole body of the ice-crystal. The pattern is based on a hexagonal or six-pointed star arrangement; and the ice crystals often reveal in their external form the characteristics which were present in the single unit. This is perfectly shown in the beautiful starlike snowflakes of cold latitudes; even, in less perfect fashion, in the frost figures on the window-pane (Fig. 10, opposite p. 57). Even the ordinary commercial ice can be made to show its elements of construction, but only by special treatment. When a strong pencil of light and heat is made to cross a block of ice, and the heat begins to break the ice-links and set the water molecules free, the cavities which are formed have the outlines of six-pointed stars; the process of undoing repeats that of construction in reverse order. The experiment is one of the most beautiful that the laboratory can show (Fig. 11, opposite p. 57).

Underlying, then, the beautiful external form of the crystal, and actually the cause of it, is a minute unit of pattern, and the hexagonal form of this unit is repeated and manifested in the hexagonal characteristics of the crystal. This was suspected before its actual demonstration in recent times, but as long as the argument had to be based on what the eye could see, no great progress was made. Of course the perfection of the crystal form of ice or any other substance could be seen to indicate a regular addition of unit to unit, extending the crystal in all directions. Inferences as to the form of the unit could be drawn from observations of the whole crystal. Just so a pavement of tiles, each rectangular in form, would naturally be bounded by faces at right angles to one another; if the single tile were triangular or hexagonal angles of 120° would be more likely. The perfection of the crystal shows that the plan of the compilation is followed strictly. A good crystal is bounded by faces of great regularity and smoothness, and the faces meet at angles which are always the same, no matter how, when, or where the crystal was formed. The faces are often large enough to behave like little mirrors.

We know now that our familiar crystals are exceptional only in their size. All solid substances tend to the perfect arrangement of the crystal, though in the vast majority of cases with only partial success. They may form, not one large crystal which we recognize as such, but a jumble of tiny crystals that escape the naked eye, and may be invisible even under the microscope. Even further the number of molecules in a single group may be so small that the term crystalline is barely deserved.

The whole of this wonderful design is on far too fine a scale to be perceived by the eye, even when helped by the microscope. The radiation which we call light, by means of which our eyes function, is too gross in texture to be of any use. One might as well use a yard measure to plan the details of the scales on a butterfly's wing.

Here the X-rays come to our aid, being radiation of the same character as visible light, but thousands of times finer in construction. They can detect the fine structure of the crystal. They are affected by the single minute unit of pattern which generally consists of a very few molecules. Ordinary light takes, so to speak, no notice of so small a body. The effect of a single unit on the X-rays is excessively small, but there are so many units in the crystal, all exactly alike and pointing exactly in the same way, that on the whole there is a result which we can observe.

It is unnecessary, and indeed it would be impossible in the time at my disposal, to describe the technique of these experiments. I refer to them only because I am anxious to show you a single molecule, just as I have tried to make evident a single atom, a single electron, and a single energy quantum. The result of the examination of any molecule can be shown as a shadow picture; as if the molecule were made to cast its shadow upon a screen. Yet it is more than a silhouette. The shadow is graded, the denser parts being cast by the thicker parts of the molecule. In the picture of the durene molecule here given, the variations of density are indicated by contour lines. The shadows are cast by the electron clouds that surround the nucleus of each atom. The nucleus does not appear: for reasons that I must pass over, the X-rays are not affected by it. But we may clearly put each nucleus at the centre of its electron cloud.

These shadow pictures are very exact. The most important of the contour lines are accurate to their own width in the drawing. The positions of the atoms in each molecule are clearly seen, and the relative positions of the molecules in the crystal. We may say correctly that we now see the single molecule clearly, and can measure its dimensions and dispositions.

We have now seen what we may call samples of those units which go to the making of the world. The single atom in the fog picture stands for the myriads of others, of ninety-two different kinds. The differences depend on the electric charge upon the nucleus, a positive charge which is balanced by the negative electrons that surround it. Hydrogen is a one-electron atom, carbon six, nitrogen seven, oxygen eight; uranium is the heaviest, and also contains the largest number of electrons, ninety-two. The names of the atoms are sometimes of an antiquity into which we cannot penetrate, sometimes they commemorate the authors or the conditions of their discovery.

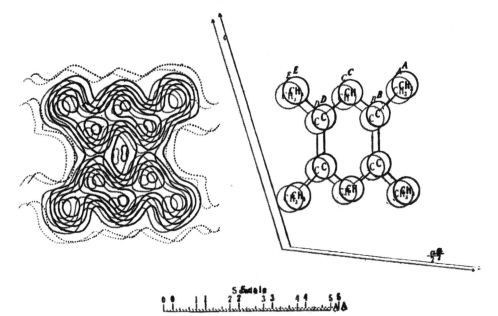

FIG. 12.—Chemical methods have shown that the skeleton of the durene molecule is the same as that of benzene (Fig. 9). Only two, however, of the carbon atoms have simple attachments of hydrogen atoms; to each of the other four, a methyl group (i.e. one carbon and two hydrogen atoms) is attached. The figure on the right represents the inference from chemical examination, that on the left is the shadow picture obtained by X-ray methods. The hexagon is regular, but is slightly distorted in the picture because, in practice, the point of view is not a matter of arbitrary choice. Two hundred and fifty million of the numbered units in the scale below the picture make up an inch (J. M. Robertson).

But Nature is in this respect less picturesque. She knows them by number only, thus affording yet another example of the tendency to order and regularity. The nuclear charge determines the power of entering into or withdrawing from combination with other atoms, and is therefore of first importance. The "atomic weight" is of less importance, but of great interest because it half displays and half conceals another of Nature's regularities. Thus the atomic weight of chlorine is 35·46, taking hydrogen as unity. It has recently been shown that there are two kinds of chlorine—so-called isotopes of chlorine—both acting in the same way because the nuclear charge is seventeen (positive), and seventeen electrons surround it. One is thirty-five times as heavy as hydrogen and the other thirty-seven, and the fractional 35·46 really indicates a mixture in certain proportions.

We have seen the results of the activities of single electrons. The charge of the electron is of the sense that we have been accustomed to call negative. The positive electron or positron has recently been brought to light, but as it differs from the other in sign only and not in quantity, there is only the one magnitude of electric charge.

We have at least seen where energy-quanta have gone by. The energy-content of the quantum emitted or absorbed in the form of radiation depends on the quality of that radiation. In blue light it is twice as great as in red light. In the case of X-rays it is far higher; and radioactive substances emit others which are more energetic still.

I have shown you pictures which give the

details of the sizes and forms and relative dispositions of the atoms in single molecules. They show also how the molecules are arranged in the solid body, with an order which, if persisted in to a sufficient extent, gives the body the regular form of the crystal. You have observed two steps in the execution of Nature's designs.

Here are we then in a world formed of materials, of which we also are formed, so that all that we do, all even that we think and say, and all our relations to the world and to one another depend upon the composition and arrangement of certain elements, limited in number, precisely defined, related to one another by curiously simple numerical laws. And we are just learning how to observe the first associations of those elements into units of pattern which contain within them the characteristics of the final assembly.

It is especially interesting to observe that somewhere in the region which we can now study life itself enters in. The life principle can become active at a point in the assembly of which we know nothing at present. We have, indeed, vague ideas of limits set by dimensions and numbers of atoms involved, but we have no knowledge of the nature of the great change. It must be observed that even if we learn the details of the mechanism of life's processes it does not follow that we can see it in motion, or understand the motive power.

As we come gradually to understand and appreciate Nature's plans, and try ourselves to think and act in accordance with them, our science must grow, and the powers which science gives us. Physics and chemistry have already made great strides, largely on account of their acceptance of the principle of the "particulate" in the case of matter. As this principle has recently been extended to electricity and energy, we may be sure that the pace of our advance will not be diminished. More and more do the sciences, and the applications of the sciences, depend upon reasonings and experiments based on the elementary details of the construction of the world. It cannot often happen that the discovery of the form and characteristics of a molecule will explain a biological observation or start a new industry. What does happen, and this is fundamental and vastly important, is that the more closely our thoughts are in accord with Nature's way of acting the more quickly we realize the significance of every suggestive observation, and the more skilfully and effectively we take advantage of it. The attitude of mind is what matters most of all. It is this awareness that has led and is leading to all the extraordinarily effective applications of physics and chemistry and the other sciences.

As I said at the beginning of this lecture, it has been suggested to me that I should say something of the progress of the applications of physical science. Obviously it is impossible for me to enter fully into an account of so wide a subject: and surely it is in some respects unnecessary. Sir James Jeans told you two days ago, in the first lecture of this series, of the marvellous excursions into astronomical space and time to which recent physical research has largely contributed. It is not too much, to say that all the other sciences have been stirred by the same impulse. The philosopher has been deeply interested. Anyone can see for himself, or can read what has been told by others of the extent to which our industries, especially our modern industries, depend upon our scientific knowledge. He has but to observe, let us say, the immense developments of electrical engineering, or of the chemical industries, or of the means of transport, or of means of inter-communication, or of the prevention and cure of disease, or of the use of materials such as rubber, alloy steels, cellulose, and so on. If one were to set to work to compile an estimate of the extent to which our various businesses now depend on science, whole volumes would be needed. It is not too much to say that the successful industries of to-day are based directly upon the science laboratory, and maintain a close connection therewith.

I shall not, therefore, attempt a survey of established facts. It would, I think, be more useful, and I hope more interesting, to my audience if I spoke of a few cases in which the connection between pure science and its application is just in the making: cases in which a way seems to be opening up between the things with which we are familiar, and the secrets of Nature's constructions. The way is narrow now, not too well marked, and trodden only by a few, but it may be that before long it will be broadened to a high way. Recent determinations of the form of the protein molecule supply a good example.

Proteins are bodies of complicated composition and structure which play a fundamental part in animal life. They are essential constituents of muscle and nerve, skin and hair and wool: the values of our foods are related to protein-content, since the body's store must be continually replenished. We have long known that carbon, nitrogen, oxygen, hydrogen, sulphur, and other atoms are elements of the protein molecule, and that its molecular structure must in some ways be very variable, whence the name. Only recently we have acquired new information about the general plan, mainly from our X-ray studies, and it is extremely curious and interesting. Every protein molecule has a backbone or central framework, consisting of an atomic chain in which carbon and nitrogen atoms recur with perfect regularity, two of the former to one of the latter (Fig. 21). The chain cannot be pulled into a straight line because the two links that join each atom to its neighbours in the chain make with each other an invariable angle. This angle, which can be measured with great exact-

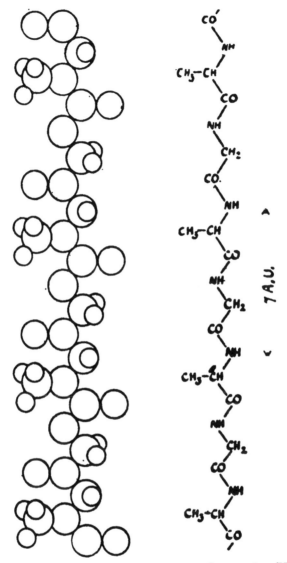

FIG. 13.—The atomic arrangement as shown by X-rays (left) and the conventional formula (right) of a part of one form of a protein molecule, viz., the fibroin molecule of natural silk. (See Astbury, *Fundamentals of Fibre Structure*, p. 124.)

ness, as can also the distance between each pair of atoms, is rather more than a right angle. Although there is this element of rigidity the chain as a whole can be crumpled up because any part of the chain can turn round the link that connects it to the remainder, as if the link were an axle. It is the shortening of the protein chains in this way which constitutes the contraction of our muscles, or the shrinkage of a textile fabric, All animal movements are produced by the contraction and extension of muscular fibre, and directed by way of nerves, in which also the protein is the important constituent.

The great variety of protein is due to variety in the nature of pendant atoms or groups of atoms which are attached to the separate links. Sometimes these are comparatively simple, as in the case of natural silk; sometimes very complicated and variable, as in the many types and conditions of hair and wool. There is a certain pendant called cystine, which contains sulphur, and is found in hair but not in muscle: it would appear that this constituent gives to hair its relatively imperishable character.

Knowledge of this kind throws light on the work of the biochemist and physiologist who try to understand the behaviour of the living body and pass on their knowledge to those whose business it is to care for animal life in any form, or, again, to others who are concerned with the supply of the nation's food. Any business which deals with animal products is necessarily interested. It is possible, for example, to cite cases where such fundamental knowledge has been of value in the textile trade. Yet specific instances do not satisfactorily illustrate the main point, which is, that understanding of Nature's ways renders observation suggestive.

It is curious that vegetable life runs parallel with animal life to this extent that here also the long chain molecule is Nature's favourite element of construction. In this case, however, the separate links are rings formed of five atoms of carbon and one of oxygen (Fig. 14). We call the structure cellulose. It enters prominently into the structure of all things that are rooted in the ground. We have, in this case, also learnt much in recent years of the molecular plan.

The preference for the long chain where life and growth are to be fostered is a very interesting phenomenon. The chain may be increased in length by the addition of new links; it is flexible; it can be extended or contracted; it has special values in one direction, endowing it with directive purpose. All these, to various degrees, are essential to that in which life is to exist, and growth is to take place.

These subjects to which I have specially referred have been respectively of the animal and the vegetable world. Let me finally refer to the mineral. I would take one instance from our knowledge of the metals. Our recently acquired knowledge of atomic arrangement in the metals and their alloys has been exceptionally full and accurate. No knowledge could have been more welcome to the metallurgist on whom demands

are now made for the supply of materials of new
and extraordinary qualities. Every variety in the

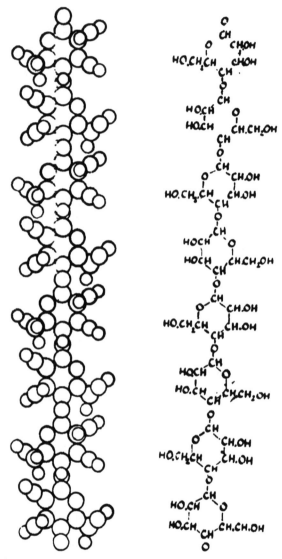

Fig. 14.—The atomic arrangement and conventional plan of
the cellulose molecule. (Astbury, *Fundamentals of
Fibre Structure*, p. 110.)

behaviour of an alloy can be traced to the arrangement of the atoms in the unit of pattern of its
crystals. Of possible combinations to form alloys

there is no end; and it can be readily understood that knowledge of structure is of the highest value in planning the course of research. Such knowledge is growing rapidly. We in this country have compelling reasons for developing to the utmost our knowledge and practice of metallurgy.

Lastly, the bulk of the earth's crust is composed of the silicates, whose molecules are formed mainly of oxygen, the most frequent of the atoms in the world, and of silicon, which is next in order of frequency. The great variety of the silicates has been a perplexing question. We have now learnt that there are fundamental rules for the combination of oxygen and silicon which in one form or another are never departed from, and we know the exact form of those rules. The great variety is due to the fact that all sorts of strange atoms can be added to the main structure in many ways. The clearer view of the highly important silicate class is important to the geologist and the mineralogist. The diagram of Fig. 15 will serve as an illustration of the arrangement of the atoms in the molecules of certain well-known minerals.

Sir James Jeans has told you that the study of astronomical events cannot be used as an argument for a mechanistic theory of the universe. Nor can the knowledge of Nature's constructions, of which I have tried to draw a rough picture this afternoon. It is true that Nature's operations move with machine-like precision, and that all her processes, whenever we are able to repeat them, follow the rules of the experimental laboratory. But it is also true that we have another

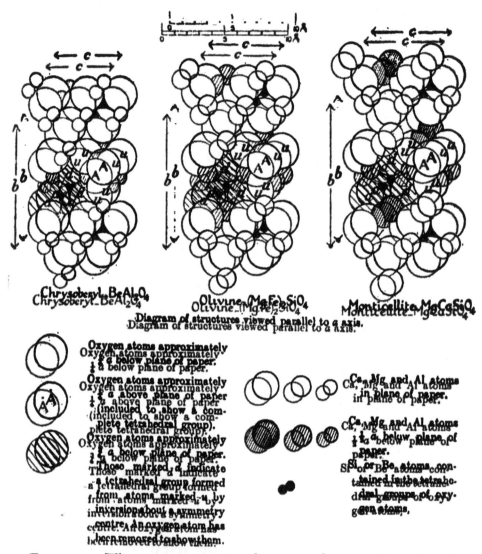

Chrysoberyl. BeAl₂O₄

Olivine. (MgFe)₂SiO₄

Monticellite MgCaSiO₄

Diagram of structures viewed parallel to a axis.

Oxygen atoms approximately ½ a below plane of paper.

Oxygen atoms approximately ½ a above plane of paper. (included to show a complete tetrahedral group).

Oxygen atoms approximately ½ a below plane of paper. Those marked d indicate a tetrahedral group formed from atoms marked u, by inversion about a symmetry centre. An oxygen atom has been removed to show them.

Ca, Mg and Al atoms in plane of paper.

Ca, Mg and Al atoms ½ a below plane of paper.

Si or Be atoms contained in the tetrahedral groups of oxygen atoms.

Fig. 15.—The arrangement of atoms in certain minerals (W. L. Bragg and J. West).

laboratory, wherever we meet our fellow men and that there also we learn by experience, and make observations on which we base thoughts and actions. We feel that we have some control over that we do, and may act selfishly or unselfishly. If the lessons of the two laboratories seem to contradict each other, the clash is not even so definite as that which in the physics world may set the wave theory and the particle theory in apparent contradiction, if we confuse the uses to which the two theories may be put. We know that in such a case the deficiency must be in our own minds, which are unable, at present, to interpret in full what we observe. Do not let us therefore be oppressed by unnecessary fears that we are but helpless cogs in a machine, but let us throw ourselves eagerly into the task of trying to interpret and live in the world in which we find ourselves.

III

PROFESSOR E. V. APPLETON

F.R.S.

CHAPTER III

THE ELECTRICITY IN THE ATMOSPHERE

THE previous two lectures have dealt with the progress of theoretical and atomic physics, and so have been largely concerned with the mathematical researches of the study and the experimental investigations of the laboratory. As a change, I invite you this evening to consider the physics of the open air, for I have taken as the subject on which I am to report progress, "Electricity in the Atmosphere." The subject is relatively an old one, and yet some of the problems connected with it have only recently been satisfactorily solved. I shall deal first with the electricity in the lower atmosphere, which can be investigated directly, and afterwards with the upper atmosphere, which is out of human reach, and which must be explored by indirect means.

Electricity in the Lower Atmosphere.—We are all familiar enough with the thunderstorm, which is the most striking manifestation of the electricity in the lower atmosphere, but even on a day free from storminess there is a good deal of electricity about. In fine weather both the surface of the earth and the air itself are electrified, and it is the maintenance of this fine weather electricity which has been one of the chief mysteries of the subject. On a fine day the earth's surface has a negative charge, while the lower atmosphere possesses, on the whole, a positive charge. Because of the attraction between opposite charges there is a transport of positive electricity from the atmosphere into the ground; and, although this downward current is so small as to require delicate instruments for its detection when we consider, say, one square yard of the ground, the total influx of positive electricity for the whole of the earth's surface is of the order of one thousand amperes.

Observation shows that the downward positive current into the ground is sufficient to neutralize the earth's surface charge in a few minutes, but no such neutralization is found to take place. The earth's negative surface charge persists. Evidently it must be replenished in some way, but the persistence is at first puzzling. The situation is somewhat similar to that of a man who can go on cashing cheques without reducing his bank balance. Clearly, in such a case, one must suppose that other cheques are being paid in at one or other of the branches, even though the accountee does not know his benefactor. Many suggestions have been put forward concerning the manner of compensation in the electrical problem, but only recently has the real benefactor been identified.

Hourly Variation of the Earth's Electric Charge. —For further insight into the problem we must consider first how the earth's surface charge varies throughout the day. Measurements of this quantity have been made now for many years. One method used is to isolate a portion of the ground and measure the charge on it, but that

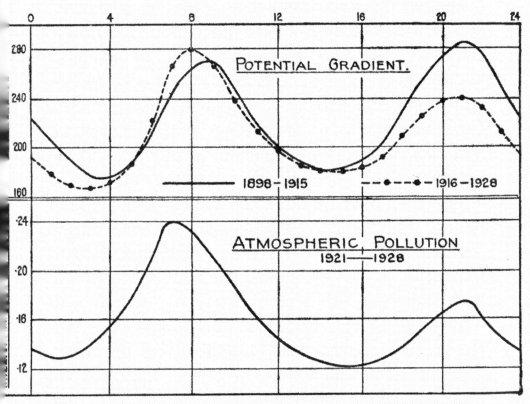

FIG. 1.—Showing hourly variation of potential gradient (above); and of atmospheric pollution in summer (below). (Kew Observatory.)

Reproduced by kind permission of Dr. F. J. W. Whipple and of the Royal Meteorological Society.

most commonly used is to calculate the magnitude of the charge by the electric force it exerts in the immediate neighbourhood of the ground.

In the first diagram, which is due to Dr. F. J. W. Whipple, you will see the results of a long series of measurements of this electric force, or the potential gradient, as it is termed, at different times of the day. The graph in the upper half of the diagram shows how the charge (which is measured by way of the potential gradient) varies with the hour of the day. You will notice an early morning maximum and an evening maximum also. The question to be decided is whether these effects are characteristic of the earth as a whole, or are merely local phenomena. The lower diagram suggests plainly that these variations are connected with the effects of smoke and dust in the atmosphere, and we now believe that these effects mask the natural one we are trying to discover. This striking parallelism between potential gradient and atmospheric pollution was first recognized by Chree and Watson in 1923. The top diagram shows also how the earth's charge was influenced by the introduction of summer time in 1916. You will see that the effect of the smoke is noticed one hour earlier than previously, meaning, I suppose, that people lit their fires one hour earlier.

I think we may say that it was due to the fact that so much attention was directed to measurements of the earth's charge on land that the real nature of the problem was for so long obscure. For, as we have seen, measurements made near cities are vitiated by the influence of smoke and dust, so that the correct variation of the earth's charge as a whole is not shown.

The first measurements which showed the natural variation of the earth's charge throughout the day were those carried out over the oceans by workers in the American ship *Carnegie*. This vessel was made of non-magnetic material, and between 1909 and 1929 made several world-wide cruises, in which surveys of the earth's electric and magnetic properties were conducted. You will probably remember that she was unfortunately destroyed by an explosion and fire in Apia Harbour in 1929. As a result of one of these expeditions it was found by Dr. S. J. Maunchley that the variations of electric charge on the earth at sites in the Atlantic and Pacific Oceans were quite different when each was studied with reference to local time, but that when the results were all referred to Greenwich mean time there was a very marked correspondence. I should add that this does not, of course, imply that Greenwich mean time is any better than any other time, but it does mean that the charges over the greater part of the world increase and decrease in the same way at the same universal time. By going away from smoke and dust we have eliminated local influences, and can see how the earth's charge as a whole behaves. Maunchley's results showed that the earth's charge increases throughout the day, reaching a maximum at about 7 p.m. G.M.T., and that it is at its minimum in the early morning. Any theory put forward to explain the mystery of the replenishment of the earth's electric charge should, to be satisfactory, explain why the charge varies in this simple fashion.

The Influence of Thunderstorms.—I need not trouble you with an account of the early theories of the maintenance of the earth's charge. They all appeared plausible at first, but one by one they have had to be discarded because of advances in our knowledge. I will proceed straight to an account of the theory which, to my mind, is the correct one, though I ought to add that there are still a few workers in this field who do not yet accept it. This theory was put forward by Professor C. T. R. Wilson, of Cambridge, the most eminent of all workers on atmospheric electricity. His theory is, in brief, that thunderstorms are responsible for the maintenance of the earth's negative charge in spite of the dissipative influence of the air-to-earth current.

You will readily see that we must apply two tests in the examination of any theory of this kind. First, it must be clear that the agency suggested should cause the return of electricity of the correct sign, that is to say, negative electricity, to the ground. Secondly, it must be demonstrated that the agency is potent enough to supply the charge for the whole earth. I will take these two points in order.

It has been known for a long time that a thundercloud must consist of two oppositely charged regions of electricity, one positive and one negative. But not till recently has it been clear how these charges were disposed. You will see that if the negative charge is below the positive charge it will tend to send negative electricity into the ground, whereas if the positive charge were underneath, the reverse effect would take place. In testing the thunderstorm theory it is therefore important to know how the two

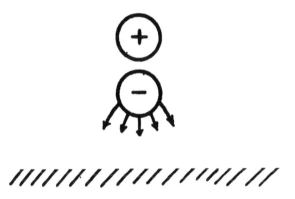

FIG. 2.—Showing diagrammatically disposition of electrical charges in typical thundercloud.

charges are situated. Now experiments carried out in this country and also in South Africa (where there is plenty of observational material in the way of thunderstorms) have proved very definitely that thundercloud formation usually results in the elevation of a positive charge above a negative one. This is shown diagrammatically in the second figure, from which it will be seen that the lower charge tends to send negative electricity into the ground.

It was pointed out by Professor Wilson that

the entry of negative charge to the earth under
a thunderstorm was likely to be specially marked
at the tips of all pointed bodies, such as the leaves
of trees and blades of grass. In order to test this
matter, Dr. T. W. Wormell in Cambridge and
Dr. B. F. J. Schonland in South Africa have made
measurements of the sign and magnitude of the
charges passing into pointed conductors in
thundery weather. Dr. Schonland used a thorn
bush as his electrical collector. The bush was
insulated electrically and the current passing into
it measured. This current was found to be chiefly
negative in thundery weather, and it was possible
to show that it was sufficiently intense to compen-
sate for the ordinary air-earth current over quite
a considerable tract of fine-weather area. Both
Wormell and Schonland, in fact, after making
a kind of profit and loss account for the earth's
electric charge in the area each has studied, are
of opinion that the thunderstorm agency is
sufficient to cause the maintenance of the earth's
negative charge.

There are really far more thunderstorms
occurring at any given time than we ordinarily
imagine. Dr. C. E. P. Brooks, from an analysis
of the world's weather records, estimates that
there are over one thousand thunderstorms in
action at any given time, so that if each thunderstorm area sends an ampere of negative current
into the ground, as certainly seems possible, there
is full compensation for the thousand amperes

of positive current which I said had been estimated as passing into the ground in fine weather regions.

But we must still bear in mind the result of Maunchley, which showed us that the earth's negative charge increased steadily during the day and decreased during the night. Can this fact be explained in terms of the thunderstorm theory? It is well known that thunderstorms occur most frequently over land areas, being comparatively rare over the sea. It is also known that thunderstorms are most frequent at any given place in the afternoon about 4.0 p.m. Now there are three very thundery regions on the earth's surface, namely, Dutch East India, South Africa, and South America. While it is 4.0 p.m. local time over these regions we should expect the earth's negative charge to be increasing. A simple examination of a map of the world shows that this period corresponds roughly from 4.0 in the morning to 8.0 p.m. at Greenwich, which is just about the period when Maunchley's observations show that the earth's charge is increasing. There is therefore good correspondence. The earth's charge increases while the sun is passing over the thundery land areas, and decreases while it is passing over the Pacific Ocean. I think that this is one of the most striking results favouring the theory of Wilson.

As I mentioned earlier, the atmosphere immediately above the ground possesses in fine weather a resultant positive charge, although there are really present elementary electric particles, or ions as they are called, of both signs. We know that the splitting up of air molecules into these elementary charges is effected by two agencies. First there is the influence of small quantities of radium and thorium, percolating from the ground into the lower atmosphere, and secondly there is the effect of those mysterious particles known as cosmic rays, which enter the earth's atmosphere from outside.

Electricity in the Higher Atmosphere.—I now turn to consider the second half of my subject, the electricity in the upper atmosphere. Here it is impossible to make experiments *in situ*, and we have to rely on such natural manifestations as are observed on the ground, together with the methods of exploration by radio waves. For example, we have, in polar regions, the aurorae or northern lights, which clearly must be electrical discharges, so similar are they to the discharges observed at low pressure in the laboratory. The study of terrestrial magnetism quite early showed that intense electric currents must flow in the upper atmosphere, especially during periods of magnetic storminess. But the most direct evidence that the higher atmosphere is electrically conducting has come from experiments with wireless waves. The chief reason for this is that a radio exploration of the upper atmosphere can be made at any time, and it is not necessary to wait for natural sequences and irregularities.

Wireless methods of investigating the electrified regions of the upper atmosphere, or the ionosphere, as it is now commonly called, are really very simple. Brief radio signals are sent out by a station, travel upwards to the ionosphere, where they are reflected, and are afterwards received again at the ground. This process takes place, of course, very quickly, for the waves travel with the speed of light, but the timing of the up and down journey presents no technical difficulties, even though it is necessary to estimate the time to $\frac{1}{100,000}$ of a second. The earlier experiments of this kind showed that the density of the electricity at levels of over sixty miles above the ground was about a thousand times that near ground-level, and it seemed clear that it could not be due to the same causes.

The Reflection of Radio Waves.—I ought here to interpolate that the existence of this reflecting ceiling in the atmosphere has important consequences in practical radio communication, for it is by way of it that wireless waves are constrained to travel to the Antipodes. Without it they would fly off wastefully into space. But the echoes which are returned from it are sometimes a nuisance especially in short-distance communication, for it means that you get your signals all doubled. You receive them via the direct path along the ground, and also via the echo received a little later from the reflecting ceiling in the upper atmosphere. Some time ago the B.B.C. sent out television signals by the Baird process, using wave-lengths which gave pronounced echoes at about one hundred miles from London, and it seemed to me that the echo-effect should be shown up as a kind of ghost-image. Through the kind offices of the *Wireless World* I was able to ask amateurs receiving these television signals to look out for ghost-images. The first response I obtained was from an amateur in Bristol, Mr. W. B. Weber, who had been very successful in receiving the Baird transmissions. He sent me a most interesting communication, saying:

1. That he had received echo images by night and not by day.
2. That the echo image was always above the main one.
3. That so far as he could measure, the shift upwards was about one-seventh of the whole picture.

From Mr. Weber's observations it was possible to calculate that the echo, or ghost-image, must have been received $\frac{1}{3,000}$ second after the main signal, and with the aid of a little simple geometry one can show that the echo must have been reflected at a level of about sixty miles, which is just the height of the Kennelly-Heaviside layer.

Echoes are sometimes troublesome in the case of transmission over longer distances. The Marconi Company have developed a very efficient system of transmission by means of which line drawings can be transmitted by wireless over the Atlantic. Here the influence of multiple echoes is sometimes to be noted. The first signal to arrive is that which has been once reflected by the ionosphere, but this is often followed by waves which have been reflected twice or three times. They make longer journeys and so arrive late. The result of this is that the transmitted type-script or picture appears doubled or trebled at the receiving station.

Measurements of Density of Upper-atmosphere Electrification.—But to-night I want to stress neither the advantages nor disadvantages of the ionosphere in radio communication. I want us to consider the results which have accrued from the use of radio waves in the investigation of the ionosphere itself. This exploration is usually carried out in a local kind of experiment, the waves being projected vertically upwards at the sender, and the echoes examined at a receiving station quite close. For example, daily experiments of this type are carried out between the Strand and Hampstead, the sender being an aerial on the roof of King's College and the receiver being situated at the Halley-Stewart Laboratory.

Now the most productive method of using radio waves for exploring the ionosphere is that involving the use of a number of wave-lengths. Short waves are more penetrating than long waves, and if, in any experiment, the wave-length is gradually shortened it is found that a value is reached where reflection ceases and the waves break through. We call this the penetration wave-length. This penetration wave-length is all-important, for it gives us a measure of the density of the electricity in the region penetrated. Measurements designed to test how the electric density varies with the hour of the day and with the season of the year have now been in progress for some years, both here and in America. These are always carried out by finding the penetration wave-length.

Since there are two main regions in the iono-sphere, there is a penetration wave-length for each, from the determination of which we can obtain a value for the appropriate electron density. In the third figure the variations of the noon values of the electron densities for both regions throughout the year are shown. It will be seen that the lower region, which we term Region E, or the Kennelly-Heaviside layer, behaves much as we should expect, being denser on a summer noon than on a winter noon. This is due, of course, to the more direct action of the sun. But when we consider the densest region of the ionosphere, we find a curious seasonal relation. The maximum effect is not found to occur in summer, but in October and November. There is also a subsidiary maximum in early spring. This curious behaviour of the top layer of the ionosphere has been much discussed.

Either the measurements must be wrong, or there must be a pronounced difference in the higher atmosphere between summer and winter;

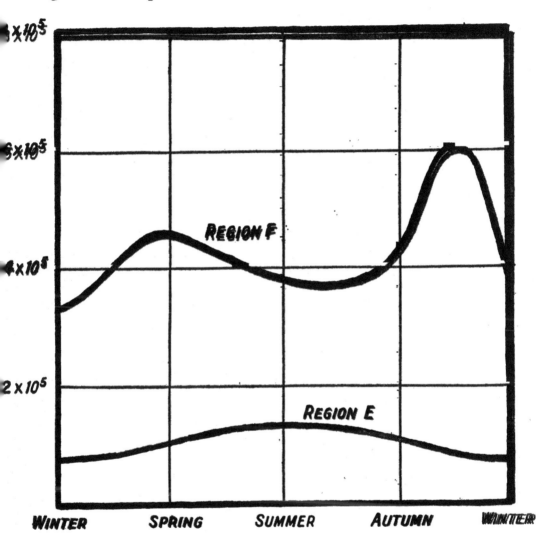

FIG. 3.—Showing variation of noon ionization for lower and for upper regions throughout the year.

for I don't think it is sound to assume that the ionizing rays from the sun differ in quality or strength from summer to winter.

Some workers (for example, in America) have concluded that the measurements are spurious, and that the electric density of the upper region is really greater in summer than in winter, the annual variation being similar to that for Region E. In this country we believe the experimental results to be sound, and to indicate that the higher atmosphere is much hotter in summer than in winter. We think that in summer the atmosphere at high levels is heated and so expands. As a result of the expansion the density of the electricity is reduced. Putting these ideas in quantitative form, it appears necessary to assume that the temperature at a level of one hundred and fifty to two hundred miles reaches a temperature of about 2,000°F. on a summer noon.

The marked seasonal variation we have just been discussing and the way in which the ionization builds up every morning at sunrise indicates very definitely that the electrification is caused by some form of radiation from the sun. Moreover, it is clear that the radiation must travel in straight lines. Now such radiation might consist of electromagnetic waves (such as, for example, ultra-violet light), or it might consist of uncharged atoms travelling with high speeds. To decide between these two possible agencies was not easy, and it was necessary to make special observations on the occasion of a solar eclipse to test the matter decisively. The result, however, demonstrated conclusively that the normal ionizing agency for both of the main regions is ultra-violet light from the sun, and that the effect of neutral atoms, if present, can only be small.

Abnormal Radio Transmission.—So far I have been concerned with what we may call the normal behaviour of the ionosphere. But it sometimes occurs that conditions are abnormal, when, for example, long-distance signalling is difficult because the intensity of the received waves is weak. At such times it is clear that the structure of the ionosphere must be different from normal. It was found, in 1927, that such periods when communication is difficult are connected in some way with magnetic storms. This connection was studied in greater detail by the British Radio Expedition to Tromsø during the Second International Polar Year, 1932–3. As Tromsø lies in latitude 70°, well within the Arctic Circle, and close to the zone of maximum auroral activity, it was a particularly suitable station for work of this kind, since the magnetic storms experienced there are of a much greater severity than those which occur in our latitude. The work of the Tromsø expedition showed that the normal density of electrification in the ionosphere in high latitudes is less than in temperate latitudes, which is just what we should expect according to the ultra-violet light theory of ionospheric origin. But on the days of pronounced magnetic and auroral activity a most remarkable result was obtained, for it was found that there were no wireless echoes at all, the ionosphere ceasing to reflect.

I do not think we can attribute the lack of echoes to a low density of the ionosphere, for the auroral displays indicate that during these abnormal periods there must be intense ionization. It seems as if there is ionization produced, shall we say, in the wrong place; that is, below the normal level. Such ionization might be expected to produce absorption, and cause the echoes to be so low in intensity as to be undetectable.

Many years ago a theory was put forward by Birkeland to explain the nature of auroral displays and magnetic storms, and in particular to explain why these phenomena are most intense in high latitudes near the poles. Birkeland's theory, put briefly, is that these disturbances are caused by the incidence of charged particles or atoms, originating in the sun, on the upper atmosphere. Birkeland's theory has been worked out in great mathematical detail by Størmer, and is accepted by most geophysicists. It has been shown that the influence of the earth's magnetic field is such as to cause the charged solar particles to converge towards the two polar regions, leaving the equatorial regions unaffected.

The Solar Cycle.—We now see that auroral displays, magnetic disturbance, and the occurrence of abnormal radio phenomena are all connected and according to the theory of Birkeland they are all to be attributed to the influence of charged particles from the sun. Now it is known that magnetic storms are connected in some ways with sun-spots, for, in a year of marked sun-spot activity, magnetic storms are found to be both frequent and severe. The last sun-spot maximum occurred in the year 1928, and about the latter half of 1933 there occurred the minimum. We are thus, at present, approaching another maximum in 1939, and both the magnetic and radio evidence already tell us that this is the case. We must therefore expect abnormalities to grow more frequent and troublesome.

But there is another question of great interest which will be examined with care during the next few years. Will the normal ionization in the ionosphere follow the sun-spot cycle? Or, in other words, does the intensity of the ultra-violet light from the sun alter in sympathy with the sun-spot cycle? Measurements made in this country have suggested that there is such a variation, but it is clear that the work in progress must be continued till 1939 before the exact magnitude of any such variation can be estimated with anything like accuracy.

Any such steady variation of the normal ionization would have an important effect for the practical radio-worker, for it would mean that each year he would have to alter slightly the wave-lengths used in order to secure optimum transmission results.

It is fitting, I think, that I should conclude this lecture at this point, showing you that the subject of atmospheric electricity still supplies us with unsolved problems and interesting work for the future. A certain amount of progress has been made, you will, I think, agree, by the radio exploration of regions which are inaccessible in any other way, but much yet remains to be done.

We are told that the mythical nymph, Echo, because of her incessant chattering, was deprived of speech to the extent that she could only repeat the question put to her. She could neither answer nor give information in any way. I think, however, that we can claim for our wireless echoes a little more than that. They repeat their original, it is true, but, as I have tried to show you to-night, they sometimes do it with sufficient lack of exactitude and at such significant intervals as to give away some of Nature's secrets.

IV

EDWARD MELLANBY

M.D., F.R.C.P., F.R.S.

CHAPTER IV

PROGRESS IN MEDICAL SCIENCE

In this lecture I propose to give an outline of progress made in medicine and medical science through the centuries. I have chosen this subject partly because most of my working life has been spent in studying scientific problems associated with medicine and partly because it seemed to me that, by surveying an aspect of a subject which touches the lives of all of us, better appreciation could be had of the great strides that have been made in the acquisition of biological knowledge in modern times, and especially in the present century.

It is an interesting fact that, even in the earliest days of civilization, the study of disease and its eradication greatly interested mankind. This being so, it is strange that until about a hundred years ago there had been relatively little advance in real knowledge of disease or its control. Now why is this? Is it because mankind possessed less intelligence up to a hundred years ago? I should say certainly not, for there is every reason to believe that people in the olden days were as intelligent as we are to-day. There are, I think, three main reasons for this long latent period: the first is that mankind had the wrong way, or what appears now to be the wrong way, of looking at the problem of health and disease: the second reason is that it took man a very long time to appreciate the fact that he could not get much useful knowledge of the human body unless he actually made it a subject of close and direct study both in a healthy and diseased condition: the third reason is that, until comparatively modern times, man had no appreciation of the value of the experimental method. He had first to learn the value of observation: then he had to realize that observation is not enough, but that it must be associated with experiment. Throughout all ages, speculation has been the great will-o'-the-wisp, and by contenting men's minds has prevented serious and profitable observation and investigation.

Let me illustrate the detrimental effects exerted on medical advance by these various factors. First, as regards the wrong attitude towards health and disease: we know that the Egyptians, the Assyrians, and the Babylonians, although they were interested in disease, regarded it as supernatural, and due to some sort of magic influence—an invasion of the body by some evil spirit. Disease could not be controlled unless this malignant spirit was first cast out from the body. Its entrance could be hindered or prevented by amulets and charms, but to drive it out of the body of a sick man prayers and incantations were necessary. When once driven out, the ravages it had made to the organs and tissues could be repaired by such drugs as opium, hemlock, squills, castor oil, etc. In those days astrology had a great vogue; also the art of divination flourished. In connection with the latter art, it is interesting to remember that the liver was the main object used for such purposes. The blood supply, the size and form of every portion of the liver of sacrificial animals were noted, and by this means the future was prognosticated. It is strange that hepatoscopy, which led to such careful examination of one internal organ and which held sway in the lives of these ancient people for thousands of years, should not have stimulated in them an interest in the form and structure of other organs, or impressed them with the fact that disease was a process belonging to the natural world. It is obvious that the view of disease as a supernatural phenomenon is incompatible with any advance in knowledge of its real nature.

It is an interesting fact that even in those early days the profession of medicine was in some way organized, although clearly, so far as incantations and prayers were concerned, much must have remained in the hands of the priests. We have evidence of the ethical basis of medical practice in the Code Hamnurabi, found on a monolith in Babylonia and dating from about 2000 B.C. This deals with the laws controlling religious, legal, and medical procedure, and contains a number of instructions to medical practitioners. Two of these may be quoted:

"If a doctor has treated a gentleman for a severe wound with a bronze lancet and has cured the man, or has opened an abscess of the eye for a gentleman with a bronze lancet and has cured the gentleman, he shall take ten shekels of silver."

"If the doctor has treated a gentleman for a severe wound with a lancet of bronze and has caused the gentleman to die, or has opened an abscess of the eye of a gentleman and has caused the loss of the gentleman's eye, one shall cut off his hands."

Medical work in those days must have been at least exciting!

Another great era of sterility in medicine, due again to a wrong attitude towards the human body, was during the Middle Ages, extending from the fifth to the sixteenth centuries. It followed the almost complete extermination of the Greeks and Romans. This obliteration of a civilization was due largely to the barbarian invasion, but partly to a series of devastating plague epidemics. Since all knowledge and culture throughout the Middle Ages remained in the hands of the Church, the outlook of men's minds was controlled by the clerical influence. It was considered that nothing in life was of any importance except those things bearing upon death, judgment, Heaven, and Hell. Man's soul was everything, his body nothing. So far, then, as health and disease were concerned, since they were conditions affecting the body they were of no account. The whole attitude of the civilized world was dominated by the particular Christian tenets of that time; this can be seen in the writings of distinguished men of the period. For instance, Tertullian said, "Investigation since the Gospel is no longer necessary."

Between these two long sterile periods, namely, that of the Assyrians, Babylonians, and Egyptians, and the Medieval Ages, there had passed the great period of Greek and Roman culture. As in all other intellectual and practical pursuits, this culture left its mark on medicine and medical science, and, after becoming moribund and being revived on several occasions from the time of its highest development to the present day, still exerts an influence on the outlook of medical men. Everyone is aware of the greatness of Hippocrates and his school, of Aristotle, and of Galen. The Greeks were the first to introduce real sanity into medicine. They attempted to shed the long-established teaching that disease was due to the invasion of the body by malignant spirits. For the first time in history disease was regarded as a natural process, to be studied and observed directly before there could be any hope of obtaining knowledge. They regarded a healthy mind in a healthy body as the most desirable thing in the world, and for the attainment of this insisted upon the necessity of living under the best and most natural conditions. The curative value of Nature herself was prominently taught. The ethical ideals applicable to medical practitioners as taught by Hippocrates represent even to-day one of the finest codes of behaviour that can be followed. It can at least be claimed for the Greek attitude towards the human body that it was the first in the history of man compatible with advance in medical knowledge either by observation or by investigation. Even Greek teachings were unfortunately cluttered up with speculation, which encumbered the wheels of progress. It will be remembered, for instance, that the Greeks thought that blood, phlegm, yellow bile, and black bile were the four dominant elements of the body, and that these four elements corresponded to and controlled the four well-known dispositions, sanguine, phlegmatic, choleric, and melancholic. When these elements worked in equilibrium and harmony, good health resulted; when they were out of harmony, a condition of dyscrasia brought about ill-health. It will be seen later that, with the resurrection of Greek culture at the Renaissance, these views again held sway for a long time, and had to be shattered before further advance in knowledge could be attained.

The second main reason for the long latent period in medical progress, mentioned at the beginning of my lecture, namely, the slowness of mankind to realize that a knowledge of the structure of the human body is essential before any understanding of what really constitutes disease can be attained, is also partly responsible for the lack of rapid advance that might have been expected in this great era of Greek culture. Dissection of the human body was taboo to the Assyrians, Babylonians, Egyptians, and also to the Greeks, and they were therefore incapable of getting any sound knowledge of the condition of organs and tissues of the human body either in health or disease. It is true that, after the Greek settlement in Alexandria, the study of human anatomy was established at the instigation of the Ptolemys, and it is from this date (300 B.C.) that anatomy can be considered as having its origin, but unfortunately it proved to be temporary and local. Even the lively and ever-searching efforts of Aristotle were frustrated by his inability to examine the human body, but he endeavoured to make up for this by the dissection of animals, such as monkeys and pigs He confessed, for instance, that he had never seen human kidneys. At a later period, Galen (A.D. 130–200) did much to acquire knowledge both of anatomy and physiology. He it was who established the fact that the arteries of the body contain blood and not pneuma, although he failed to discover that the blood circulates, or even that the pulsation of the heart drives the blood into the vessels. With the disappearance of Greek culture in Alexandria the study of anatomy again lapsed, and during the period of the

Middle Ages all matters of medical interest became practically moribund. It is true that the subject was kept alive to some extent in Alexandria after conquest in A.D. 640 by the Arabs, also at Byzantium and at Salernum in Southern Italy. Although very praiseworthy, these indications of activity were chiefly of interest in that they saved Greek medicine from entire destruction, but they had at the time no general or wide influence on the development of medical knowledge. It was only with the Renaissance, and especially after the establishment of the medical school of Padua, that the study of anatomy was properly revived and developed to a high art. It is impossible to consider this great and revolutionary period and its effect on medicine in any detail, but it can be said quite definitely that such men as Leonardo da Vinci, Vesalius (1537 onwards), and Fabricius (1594), completely altered our whole knowledge of the structure of the human body. With this great development in anatomical studies the teachings of Hippocrates and Galen were reborn and a new period of medical culture arose.

The lack of appreciation of the experimental method was mentioned above as the third reason for the long latent period before medical science advanced, but the time has now arrived when a classical demonstration of this method of investigation was given to the world. Harvey was a student at Padua, and was greatly influenced by the teachings of Fabricius, being particularly interested in his observations on the presence of valves in the veins. It was in consequence of this interest that Harvey, on his return to England, settled down to study the circulation of the blood. The outcome of this study was the publication in 1628 of *De Motu Cordis*, in which he describes the experimental methods whereby he proved that the blood actually circulates through the body. This publication is now recognized as not only the first but also one of the most beautiful demonstrations ever recorded of the experimental method, and shows the possibilities of this method in bringing to light new facts. It must be remembered that even in the beginning of the seventeenth century people still believed in Galen's theory (A.D. 130–200) that the blood was present in two different systems of vessels, in each of which it ebbed and flowed continuously, but did not circulate. It was many years before Harvey's view of the circulation of the blood was generally regarded as true. It might be thought that the success of Harvey's work would have resulted in the general adoption of the experimental method by other people interested in the study of disease. Unfortunately, this was not the case, for between the time of Harvey and the time that the experimental method was generally adopted as the main plan of action in biological and medical research, two hundred years had to pass. This, of course, does not mean that no experiments were done in the meantime, but the statement is in general true. The world was still almost completely ignorant of those facts relating to health and disease which could be established by pure observation apart from experiment, and there was no basic knowledge of chemistry and physics at the time to allow much advance. Chemistry itself began to take shape about the time of Harvey's life, and however much we deprecate the tardiness with which experimental biology developed, the same cannot be said of chemistry and physics. The mention of the names of such great chemists as Glauber, Willis, Mayou, Agricola, and Stahl, leading to Robert Boyle, Cavendish, Priestley, Lavoisier, and Dalton, is sufficient to show that chemistry as we know it originated in and developed from this time. There were, of course, numbers of distinguished physicians during the seventeenth and eighteenth centuries, but their interest in disease was rather as observers of its Natural History. Sydenham (1624–89) represents Hippocrates at his best, and there was also the distinguished Dutch physician, Boerhaave (1714), who introduced sane methods involving accurate observation into the study and teaching of medicine. The drawback of this period, however, was that even the most distinguished physicians, instead of studying and advancing medical knowledge, were intent on inventing hypotheses based on pure speculation to account for all disease. Two of the best-known so-called systems of medicine dating from this time were those of Cullen, who regarded all disease as being due either to spasm or atony, and of John Brown, who put forward the hypothesis of excitability as the basis of disease, a hypothesis which caused unlimited discussion and controversy among those interested in medical problems. So far as medicine was concerned, these discussions were fruitless and a great waste of time.

In this relatively sterile period of the eighteenth century the most outstanding advance came again from Italy in the work of Morgagni (1760), who, by his accurate post-mortem examinations succeeded in establishing the subject of pathological anatomy. When we remember how dependent knowledge of disease is on discovery of the condition of organs as found *post mortem*, we can realize what a revolution in the study of medicine was made by this distinguished investigator. Rather later, under the stimulus of John Hunter (1728–93), pathology again made a great advance. Hunter was a naturalist keenly interested in pathology, and his very active life was spent in establishing closer union between medicine and natural science. He it was who wrote to Edward Jenner, "Don't think,—try," when the latter was interesting himself in the observation that dairymaids who had contracted cowpox in the course of their work were immune to smallpox. The result of this admonition on

Jenner had, as everyone knows, a revolutionary effect on the history of smallpox, for as the outcome the system of vaccination was developed, and its efficacy in preventing smallpox was established.

In this period, largely through a group of distinguished French physicians, physical examination of the body both in health and disease became much more accurate. In 1819 Laennec introduced the stethoscope, and from this time onwards it was possible for medical men to use both auscultation and percussion in diagnosis. It was this group of physicians, also, who extended the work of Morgagni in the *post mortem* room to an accurate study of the cases prior to death. There was, in fact, at this time great progress made in the clinical diagnosis of individual diseases. Indeed, the material upon which medical research could then work was rapidly accumulating, although even now the experimental method itself was dormant.

Modern medicine can be said to date from the time of Pasteur and Claude Bernard, a time when the experimental method was finally adopted as the chief instrument of medical science. In the years 1857 to 1860 Pasteur published his classical papers on lactic acid and alcoholic fermentation. As the result of this work Pasteur was imbued with the idea, which later he did a great deal to establish, that infectious fevers were due to the spread of a *contagium vivum*. This view he developed because he was convinced that the process of fermentation and the process of infection were closely related. Pasteur's work led at once to the investigations of Lister, which established the fact that infection of wounds was due to the presence of germs of various low forms of life, and, following upon this, Lister developed the method of preventing sepsis in wounds by chemical substances having antiseptic properties. Thus Lister revolutionized surgery and made it practicable and safe. Another direct outcome of Pasteur's investigations was the life-work of Koch, who in 1876 published an account of his researches on the aetiology of anthrax. It will be remembered that Koch grew the anthrax bacillus from cases of anthrax and by reinoculation into animals produced the disease in them. In 1882 he also discovered that tuberculosis was caused by the tubercle bacillus. Koch was the first to grow pure cultures of micro-organisms, thus establishing the subject of bacteriology. Before long there followed the isolation of the micro-organisms responsible for typhoid fever, diphtheria, cholera, tetanus, plague, pneumonia, and gonorrhoea.

Prior to the work of Koch another distinguished Frenchman, Claud Bernard, was using the experimental method to determine the functions of many organs of the body. He it was who, by discovering the glycogenic functions of the liver, opened up the whole subject of internal secretion.

Whereas Koch might be regarded as the father of bacteriology, it seems reasonable to ascribe to Claud Bernard the fatherhood of modern physiology.

Another subject also came to life at this time. namely, cellular pathology, through Virchow, who used the microscope to determine the real structure of the cells of different tissues and associated their appearance with specific disease. In this period of great activity, i.e. from 1850 to 1880, the modern subjects of physiology, bacteriology, morbid histology, and surgery were definitely established as medical studies.

Before turning to the medical advances of the present century, let us just briefly recapitulate the changes that have been mentioned above. We have seen medicine emerge from a period of magic and religion to a stage where disease came to be accepted as a phenomenon of nature. Following this there came a time when the structure of the body was investigated; this led to the further stage in which clinical signs and symptoms and anatomical structures of diseased organs were correlated. Ultimately, the present era arrived, when the experimental method was seriously applied to the study of the body, with the result that big strides were made not only in knowledge of the actions of many organs, but also of the causes of many diseases. There were still, however, at the beginning of the present century, many diseases which had been distinguished as entities but about which we knew little or nothing as to causation or treatment, and, although this is still the case, the work of the last thirty years has helped to fill in many blanks.

Let us first see why there has been this tremendous burst of activity in medical research since the beginning of the present century. There are a number of reasons for this, among which may be enumerated the following:

1. Success breeds success. It is impossible for anybody to see the results of work such as that of Pasteur, Lister, and Koch without receiving inspiration and stimulus therefrom. It seems reasonable to believe that the knowledge and benefits produced by their investigations can be added to by others if the same or similar methods are adopted. This wonderful spirit of optimism has justified itself and has established the belief that no problem of health or disease is too difficult to tackle, and that there are no limitations to new knowledge procurable by the experimental method.

2. A second reason is that modern research has led to great advances in the basal sciences of medicine, namely chemistry, physics, physiology, pathology, and pharmacology, many of which advances have shed new light on the healthy and diseased body and have put new weapons into medical hands for the study of disease, its diagnosis, prevention, and treatment.

3. A third reason for this increased activity

is that the whole civilized community has adopted the Greek attitude towards life, namely, that health is a great asset. Ill-health is not only uneconomic, but is largely responsible for the pain, suffering, and grief in the world. From a practical as well as an emotional point of view, therefore, everything should be done to eliminate disease either by prevention or cure, and, if investigation could lead to the necessary knowledge, such research must be fostered.

The general adoption of these views ultimately led both the State and private individuals to encourage medical research, and thereby bring new knowledge to bear upon the elimination of these defects of mankind. This movement was started in many countries throughout the world.

In England in 1913 the Government set up what is now known as the Medical Research Council, a body which, under my predecessor, the late Sir Walter Fletcher, developed into a most effective system for encouraging the study of problems of disease. The Medical Research Council has also had the privilege of working in association with a number of privately established schemes for advancing medical science. It has, for instance, worked closely together with such well-known movements as the Rockefeller Foundation, with the Lister Institute for Preventive Medicine, with the Beit Memorial Research Fellowships, and with the Sir Halley Stewart and Leverhulme Trusts.

In addition to the establishment of methods for procuring new knowledge, public action led to the inauguration of the Ministry of Health, a body which in the last twenty years has shown great development. This resulted in a large increase in public medical services throughout the country, examples of which can be seen in ante-natal clinics, child welfare centres, centres for tuberculosis and venereal disease. The main function of all these bodies is to apply the scientific facts established by investigation as soon as possible to the prevention and cure of disease, and in general to the maintenance of public health.

Let us now see some of the effects of this great outburst of medical activity as evidenced in the increase of medical research and the administrative services. Some of these results can be best seen by examination of the general mortality rates and the death-rates due to specific diseases. Diagrams showing these mortality rates are given in Figs. 1, 2, 3, 4, and 5. It will be seen that in the period 1896 to 1900, 156 infants per 1,000 births died before the age of one; whereas in 1934 this figure was reduced to 59 per 1,000. In Fig. 2 the mortality rate of men between forty-five and fifty-five years of age was 20·3 per 1,000 in the period 1870 to 1875; in the period 1926 to 1930 it was 11·7 (Fig. 2). The death-rate from tuberculosis came down from 3,478 per million in the period 1851 to 1860 to 740 per million in 1934 (Fig. 3). The death-rate from whooping cough decreased from 510 per million in the period 1871 to 1880 to 51 per million in 1934 (Fig. 4); in the case of measles from 380 per million to 93 per million (Fig. 5); in

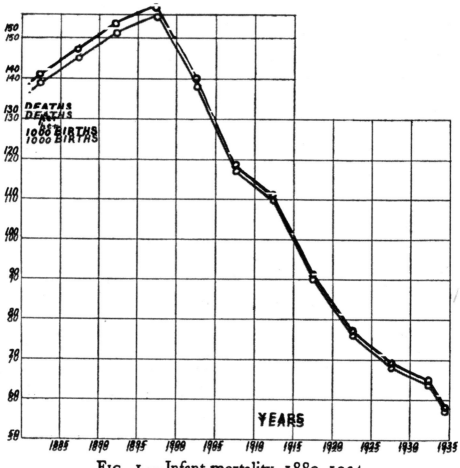

FIG. 1.—Infant mortality, 1880–1934.

the case of typhoid fever from 320 to 4 per million (Fig. 4). In most of the above diseases the actual fall in death-rate, sometimes of a startling nature, has occurred in the present century. In the case of tuberculosis, however, the fall has been going on continuously now for the past hundred years. Using these figures as a test, it is clear that there has been a tremendous improvement in the standard of health, especially in the past thirty years.

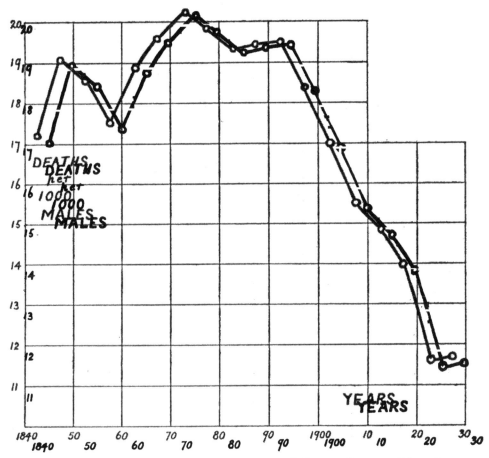

FIG. 2.—Mortality rates, males aged forty-five to fifty-five, from 1840 to 1930.

There are, however, other methods of reaching the same conclusion. Although we have obtained a great deal of additional control in the diagnosis, prevention, and treatment of disease, it is undoubted that a number of diseases have disappeared for reasons other than direct control.

For instance, there has been a virtual disappearance of some diseases for reasons which can only
be guessed. Gout, which a hundred years ago
was a very common disorder, is now relatively
rare. Why this disease has practically disappeared

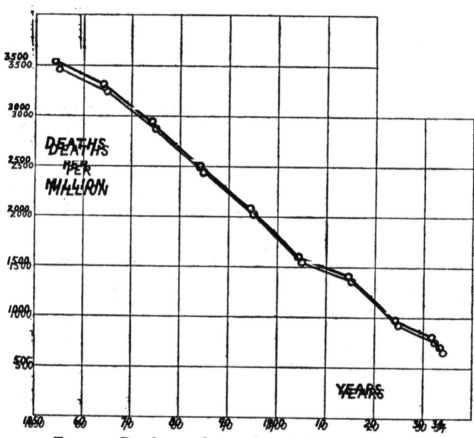

FIG. 3.—Death-rate from tuberculosis, 1850–1934.

is not known, but there are some who think its
reduction in incidence is related to the diminution
of lead poisoning. A second disease which has
practically disappeared during the past thirty
years is chlorosis, a particular form of anaemia
in women. Twenty-five years ago the out-patient

departments of hospitals were attended by large numbers of young women suffering from this disease; nowadays the average medical student hardly recognizes it. It has been suggested that

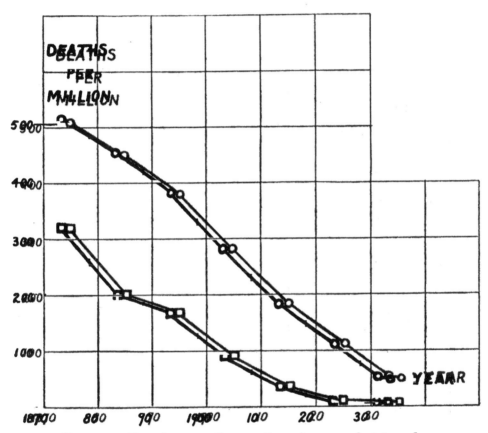

Fig. 4.—Death-rates from whooping cough (o) and typhoid fever (□).

chlorosis has disappeared owing to the discontinuance by women of tight-lacing; this, however, is only a suggestion. Another disease which used to be of a deadly nature is summer diarrhoea and vomiting of infants, a condition which became almost epidemic in those years having prolonged spells of hot weather. Many present in this room may remember the summer of 1911, when thousands of children died owing to this condition. It has been suggested that the great reduction

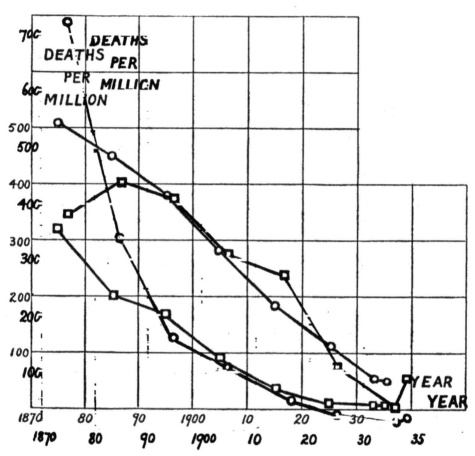

FIG. 5.—death-rates from scarlet fever (o) and measles (□), 1870–1934.

in this disease is due to the disappearance of flies from town life, a result following the replacement of horses by the motor-car. Here again, however, the explanation offered is by no means established.

Another important reason for the improvement

of health is undoubtedly the reduced consumption of alcohol by the community. There is probably more in this than we at present realize, but even now it can be stated definitely that certain diseases known to be associated with large alcohol consumption have been greatly reduced since the Central Control Board (Liquor Traffic) succeeded in making the country comparatively temperate as regards alcohol consumption; for instance, cirrhosis of the liver, which was once a common cause of death, is now much less common. Delirium tremens also has become a much rarer disease. It is probably also partly for this reason that the smothering of children in bed is now a rare incident as compared with twenty years ago.

A third reason for the disappearance of disease is the great improvement in sanitation and personal cleanliness. To England belongs the credit of taking the lead in passing laws dealing with sanitation, and this country ought not to forget what it owes to a group of distinguished men who, often against great opposition, persuaded Parliament to pass the necessary laws affecting public health. Little do we hear of such men as Chadwick, Murchison, Simon, Acland, Buchanan, and Benjamin Ward Richardson nowadays, although their great labours in the second half of the nineteenth century resulted in a dramatic decline in disease.

Efforts in Parliament were greatly assisted by cholera epidemics, which stimulated the passing

into law of the first Public Health Act of 1848
and the Sanitary Act of 1866. Organized sanitation, although started many years before the
work of Koch, and having an empirical basis up
to that time, was greatly increased when the
bacterial origin of many diseases and the fact
that they were often water- and milk-borne became
established. It is indisputable that such diseases
as cholera, malaria, plague, typhus, and typhoid
fevers have practically disappeared as the result
of improved sanitation and cleanliness. Of these,
the last to give way was typhoid fever, and, as
can be seen in Fig. 4, there are now only occasional deaths from this fever in Britain, about
four deaths per million of the population per
annum. Nowadays there is, of course, no excuse
even for this low figure.

Let us now turn to the large group of diseases
where increase in knowledge as the result of
medical research has profoundly affected diagnosis,
treatment, and prophylaxis. Roughly speaking,
additional knowledge of disease has developed
in modern times in three directions. The first
large group includes those diseases due to the
invasion of the body by protozoal, bacterial, and
virus micro-organisms. From the study of these
conditions a new branch of bacteriology known as
immunology has emerged: a knowledge of active
and passive immunity has provided medical men
with the means to modify to a greater or less
extent such bacterial diseases as diphtheria,

cerebro-spinal fever, measles, and scarlet fever. In the case of smallpox we have already seen that since the days of Jenner this could be prevented or mitigated as the result of the production of immunity following vaccination. Recent work on measles shows also that even in this virus disease much of its deadliness can be avoided. In recent years it has been shown, for instance, that immune serum obtained from a child who has recovered from measles, or even adolescent serum, when injected into a child who has been exposed to infection, can either suppress the condition or greatly reduce its virulence, according to the desires of the doctor. Unless the child is under three years of age, or for other special reasons, the medical man does not wish to suppress the measles altogether, but to allow a slight attack, which in turn induces active immunity to any subsequent attack. How excellent such treatment is can be seen in some recent work by Dr. Gunn for the London County Council. In a series of three hundred and ninety-nine children inoculated with immune serum after exposure to infection not one child died, whereas in the control group, similarly exposed but not inoculated, 5 per cent died. Recent results also obtained in scarlet fever by means of antitoxin treatment have shown how effective this treatment can be in lessening the death-rate and lowering the incidence of their sequelae.

The possibility of reducing the incidence of

middle-ear disease following measles and scarlet fever makes the modification of the course of these diseases especially important. It will be remembered that a great amount of deafness in this country is due to middle-ear disease, and until this serious condition can be suppressed the problem of deafness will remain. Methods for the substantial reduction at least of its incidence in measles and scarlet fever seem now to be available, and the general adoption of these methods is urgently needed.

As regards protozoal diseases, another form of therapy has been introduced. I refer to the branch of therapeutics known as chemotherapy. In this connection we have, of course, salvarsan and its derivatives, specific remedies for syphilis, the use of which has greatly reduced not only syphilis but also the later syphilitic nervous manifestations, including general paralysis of the insane and tabes dorsalis or locomotor ataxia. In the case of malaria, quinine, which has long been the specific remedy, has been to some extent replaced by another chemotherapeutic compound, namely, atebrine. In sleeping sickness Bayer 205 is very effective in the early stages.

The second great group of diseases brought to light in recent years is that due to abnormalities of internal secretion. We have seen, for instance, cretinism and myxoedema cured by the active principle of the thyroid gland, namely, thyroxin, diabetes mellitus controlled by insulin, tetany

controlled by parathormone, the active principle
of the parathyroid gland, Addison's disease controlled by a substance obtained from the supra-
renal cortex, and pernicious anaemia controlled
by the anti-anaemic principle of liver. While it
is impossible to discuss adequately the great
triumphs of medical science associated with this
group of diseases, attention may be drawn to the
fact that recently Dakin and West have obtained
in pure or nearly pure form the anti-anaemic
principle of liver curative of pernicious anaemia.
One of the most striking therapeutic actions that
can be seen in medicine is that which follows the
injection of 0·1 or 0·2 of a gramme of Dakin
and West's substance into a patient lying almost
moribund, suffering from pernicious anaemia.
This small quantity injected once weekly will
convert a patient within a few weeks into a
normal person with good colour, and able to live
a normal life. This branch of medical research is
not only being actively pursued, but is one of the
most promising. Anyone who has seen some of
the physiological and pathological effects produced by the sex hormones oestrin, progestin,
and androsterone, must have realized the unlimited possibilities that may result from progress
in this branch of medicine.

The third large group of diseases now amenable
to medical treatment includes those due to
nutritional defects. In the course of this work the
necessary food factors known as vitamins have

been isolated, and in some cases their constitution determined. The result is that we now know a great deal about the aetiology of such diseases as rickets, defective teeth, susceptibility to infection and its relation to diet, scurvy, beri-beri, and disorders such as pellagra, lathyrism, and convulsive ergotism. One important result of the development of this branch of medical science is that it is now accepted that diseases are not always due to the invasion of the body by a *materies morbi* agent, but are often due to the deficiency or excess of some chemical agent which is a normal and essential constituent of the body. Although it is only recently that this idea of the aetiology of disease has been established, it will be remembered by those acquainted with the history of medicine that the idea of the deficiency of a chemical substance being responsible for disease is by no means new. So long ago as 1850 Chatin advanced the view that simple goitre was due to a deficient intake of iodine, and brought forward good evidence in its support. One of the sad chapters of medicine is that Chatin's work, when repeated and examined by the French Academy of Sciences, was considered to be wrong because the members of the commission appointed did not think that such small amounts of iodine in food could ever have such a profound effect in the production or prevention of disease. It was only after 1895, when Bauman showed that the thyroid gland contained iodine, that Chatin's original observations began to receive well-deserved recognition and confirmation.

The aspect of medicine most strongly emphasized by work on the nutritional basis of disease is its prophylactic or preventive side. Although it is true that such diseases as scurvy and beri-beri can be cured by the administration of a suitable dietary, it is more important to recognize that these diseases can be prevented if proper feeding is always practised. The same is true of rickets, a disease which I will consider later at somewhat greater length.

I wish to take this opportunity of saying a few words on the subject of prophylactic and curative, remedies, as there is much wrong-thinking, both in the medical profession and among the public, as to their relative merits and importance. A curative remedy, even if only of a temporary nature, is generally dramatic in its action. The disease is there at one moment and is relieved the next, and nobody can forget the relief afforded by the treatment. In the case, however, of an effective method of preventive treatment the disease does not develop, and may be never thought of again, especially when the method of prophylaxis becomes a part of ordinary life. Now it requires but little contemplation to realize that of these two methods the second, namely, the prophylactic method, is far and away the better. By such means diseases can be completely swept away, whereas in the second case, where curative remedies are used, the best that may happen is that life is prolonged, and the signs and symptoms of the disease mitigated or cured in individual people, but the disease is not eliminated. I wish therefore to point out that in any consideration of the treatment of disease it ought to be remembered that preventive treatment is the one to be aimed at, and its very success may be measured by the degree to which the original discovery is forgotten. One of the anomalies of medical research is that, up to a point, methods of control of any disease have no relationship to knowledge of that condition. Of course, if we have sufficient knowledge of the cause of a particular disease we have at hand methods for both prevention and cure. But, short of this wider knowledge, we may be able to alleviate suffering and yet know little or nothing about the cause of the trouble, and, in other cases, we may know a great deal about a disease without having the particular knowledge to delay its progress or effect a cure. For instance, there are few diseases in the world that we know more about than cancer, yet our knowledge of methods for curing this disease is small, and of preventing it nil. On the other hand, in such diseases as diabetes and pernicious anaemia, of whose causes we have little or no knowledge, we can control their harmful effects by insulin and liver-active principle respectively, and restore the patient to a greatly improved condition.

I wish now to say a few words about a specific medical investigation, partly to illustrate how knowledge of the cause and treatment of disease can be sometimes elucidated, and partly because it was this investigation that led to the recognition that, even in this country, where such a large variety of foodstuffs is available, there is much fundamentally wrong in the feeding habits of the people and much avoidable ill-health resulting therefrom. For centuries rickets has been a scourge in this and in many other civilized countries. Although it is now relatively rare in its worst forms in London, it is still common in many industrial towns in the north. As most people well know, it is a disease in which there is defective calcification of bones to such an extent that great deformity may be produced. Up till 1914, or thereabouts, there were many views as to the cause of this disease, either of a dietetic, hygienic, infective, or internal secretory nature. The *Encyclopaedia Britannica* of 1911 ascribes rickets to the action of a toxin from the alimentary tract. Now anybody intending to find the cause of a disease of this nature would first ask himself whether the disease is known in animals, because if the condition can be produced in animals, then the experimental method can be applied, and each possible cause can be tested in turn. An experimental investigation of this kind is, of course, impossible on human beings. It has long been known that young dogs sometimes develop rickets. The first part of the inquiry, therefore, was to find conditions under which the disease could be constantly developed at will in young dogs. After much experimenting it was discovered that certain diets allowed the development of rickets, and that small changes in these diets prevented the condition.

The basis of the experimental diet was as follows:

Separated milk, 150 up to 250 c.c. daily.

Fat, 10 grammes; e.g. butter, olive oil, lard, or cod-liver oil.

Orange juice, 5 c.c.

Yeast, 5 to 10 grammes.

Cereal, such as white flour, rice, oatmeal, maize, 100 up to 200 grammes.

Lean meat, 10 to 20 grammes.

It was discovered that, according to the type of fat that was put into such a diet, rickets did or did not develop. If, for instance, an animal fat such as suet or butter, or a fish fat, such as cod-liver oil, was included, no rickets resulted; yet in other animals of the same litter receiving the same diet in which olive oil, linseed oil, peanut oil, or lard was included rickets developed. It was clear, therefore, that since the only variable in each experiment was the fat, some fats contained a substance which brought about hard bones, while other fats, especially the vegetable fats, did not contain this substance. This effective constituent was called at the time the calcifying vitamin, and is now known as vitamin D. It is the property of vitamin D to bring about the deposition of the hardening substance, namely, calcium phosphate in growing bones; when it is absent, even although there may be plenty of calcium and phosphorus in the diet, the bones will be comparatively soft and deformed.

Examination of the above dietary will show that it is possible also by altering the cereals of the diet to test their relative effects. Thus, to a litter of puppies, all the ingredients may be given in the same amount, except that one puppy may receive white flour, another oatmeal, another rice, and so on. Now, if that is done, it will be found that, according to the type of cereal eaten, in the absence of sufficient vitamin D, the severity of the rickets that develops varies with the different cereals. It will be found, for instance, that oatmeal and maize produce very bad rickets, whereas the degree of rickets produced by white flour and rice is relatively less.

These experiments show, therefore, that the development of rickets does not only depend upon a deficiency of vitamin D in the diet, but also on the amount and kind of cereal eaten. It is a surprising fact that those cereals which contain more calcium and phosphorus, such as oatmeal and maize, produce worse rickets than the cereals, such as white flour and rice, containing less of these elements, although the disease is essentially one of a deficiency of calcium and phosphorus in the bones. This result was completely unexpected, and shows how impossible it is to arrive at the truth in biology by speculation, and that the only safe thing to do in any case is to test your ideas before accepting them, even when they appear self-evident.

The above experimental work was then transferred to the study of teeth because it was noticed that the dietetic conditions which produced bad bones also produced defective teeth, and those responsible for good bone formation produced well-formed and regularly arranged teeth. From these studies we now know that such substances as milk, egg-yolk, cheese, cod-liver oil, and butter tend to produce perfect teeth, whereas the other class of substances, including oatmeal, maize, barley, bread, and rice, that is to say, the whole group of cereals, tend to produce badly formed teeth. Since the milk teeth of children, as in the case of many animals, are formed either in earliest life—i.e. *in utero*—or shortly after birth, the teaching of these experiments is clear, namely, that in order to produce perfectly formed teeth the diet must be especially rich in the calcifying foods mentioned above at that time. The permanent teeth are formed from birth till full growth, and their degree of perfection is similarly controlled by the food eaten. It is known that the teeth of the average individual in this country are defective in structure, and this must mean that as a general rule the diet of children, and even of the woman in pregnancy, is bad from the point of view of teeth and bone formation. It can be said with some certainty of prediction that until these dietary habits are altered the high incidence of dental decay and pyorrhoea will remain the scourge they are to-day, and it is only by insisting that far more milk, egg-yolk, cheese, butter, vegetables, etc., are included in early dietaries, and less cereals are given, especially in infants and children, that they will be eliminated.

I have only had time to mention two aspects of the importance of proper nutrition, but much more could be said on this subject, and, indeed, it is becoming clear that many of the physical, and possibly even mental defects so commonly found in this and other countries could be greatly reduced by the adoption of the dietetic principles above enunciated. It is this aspect of modern preventive medicine which will probably lead to a new standard of health among mankind.

I have now completed my task, and given a brief survey of the advance of medical science throughout the ages, and more particularly of the spate of knowledge that has come in the last fifty years. Criticism has been directed in this lecture to the earlier attempts of mankind to deal with disease, and the great errors both of commission and omission which prevented the advance of knowledge have been pointed out. It is interesting to surmise what people will think of our present-day efforts a thousand years hence. As a matter of fact, it is quite likely they will never think of this subject because disease will be so rare that it will require a mental effort to remember that there was a time in the world's history when practically everybody suffered from some ailment or other. If, however, some antiquarian looks back and examines our literature, he will no doubt be interested in the fact that it was in this period of the world's history that we first not only realized but applied in a practical way the idea that increase in knowledge comes rapidly by the so-called experimental method. He will probably wonder why we took so long in coming to this decision, and I have tried to give to-night some of the reasons for this long latent period. The danger is that we accept the existing situation without remembering the great effort which has been necessary on the part of mankind to arrive at our present point of view. Never again must we return to the condition where speculation without experiment holds sway.

There are many things associated with present-day health questions which will cause amusement to our antiquarian philosopher. He will no doubt, for instance, entertain his friends with a description of the methods of the medical practitioner who, either because he thinks it right or because patients demand it, supplies many people coming under his care with a large series of remedies, including a bottle of medicine, a gargle, a dusting powder, and an ointment. If, however, he laughs at this, what will he think of the present situation in this country when he sees a land flooded with that type of food which is probably more essential than any other single factor for the promotion of good health, but under such conditions that many people cannot get it? I refer especially to the position of milk in this country at the present time. When he sees, for instance, that it is possible for a manufacturer of umbrella handles and buttons to obtain large supplies of milk at the price of 5d. a gallon at a time when it is impossible to obtain the same milk for infant feeding at less than 2s. 4d. a gallon, he will no doubt wonder whether the country had any sanity about it at all. This, however, is no doubt a passing incident. It is all so foolish that it cannot possibly continue for long. When we once realize the value of proper feeding, especially in early life, the improvement in the public health will be so evident that its adoption will remain permanent just as improved sanitation and cleanliness have established themselves.

At the present time, as regards the general problem of medical investigation and the accession of new knowledge, it can be said that the pace is quickening, and facts are accumulating faster and faster, and there seems to be no limit to the possibilities of discovery which will follow the present methods of investigation. So fast, indeed, is the progress that in preventive medicine, at least, the administrative officer finds it difficult to keep pace. It would be good if we all realized that we are passing through a golden age of medical science, and I hope that one, and probably the main, outcome of this lecture will be to impress upon all of us the desirability of counting our blessings.

V

PROFESSOR J. B. S. HALDANE

F.R.S.

CHAPTER V

HUMAN GENETICS AND HUMAN IDEALS

THE object of the Halley Stewart Trust is the investigation of the application of Christian ideals in social life. There is perhaps a certain conflict between science and Christian ideals, and in some cases I think that it is a real conflict. Where that is so I can say without hesitation that I am in favour of science. But in certain cases the support of science is claimed for ideals which were there long before science, and which Christianity claims to have superseded by nobler ideals. To-day very remarkable demands are being made in the name of biology in this country, and also to a much greater extent in certain foreign countries.

I want to examine two theories which may be briefly stated as follows. The first is the theory that racial health necessitates the sterilization of the unfit. The second is the theory that some races are superior to others, whose members are incapable of rising to the highest levels possible to humanity.

These two theories are generally, but by no means always, held together. The second, I think, we may say is definitely anti-Christian. It is inconsistent with the views which St. Paul took in the First Epistle to the Corinthians. The first theory, that concerned with sterilization, has been formally condemned by the present Pope, but it is, I think, less obviously opposed to Christian ideals, and it is held by a fair number not merely of Christians, but of idealistic Christians. Perhaps they are not always aware of the consequences to which it leads.

Both of those theories are based to a considerable extent on an analogy with the principles

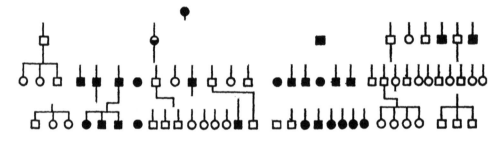

FIG. 1.—Pedigree of lobster claw (all spouses normal).

□■ Males, ○◒● Females.
□○ Normal, ■● Hands and feet split, ◒ Feet only split.

(From the *Treasury of Human Inheritance*.)

which are known to hold in the breeding of domestic animals. I shall consider later how far that analogy is applicable, and what important differences exist in the actual application of the laws of heredity to man on the one hand and to domestic animals on the other. But, before doing so, I wish to examine the evidence for the first theory, namely, that the sterilization of the unfit is necessary.

Fig. 1 is a pedigree of a congenital abnormality called lobster claw or split foot, in which the hand is reduced to an appendage with one thumb and one finger, and the foot is similarly deformed. It will be seen that this condition is transmitted, with one doubtful exception, only by affected persons, and to approximately one-half of their children, without regard to sex. In scientific terminology we say that it is due to an autosomal dominant gene. It is clear at once, from looking at a pedigree of this kind, that, if all the affected persons were sterilized, the abnormality would be wiped out in one generation, with possibly very rare exceptions due to a process called mutation, by which the abnormal gene arises anew. That process is so rare that it certainly does not happen more often than in one person per hundred millions or so. So, by this measure of sterilization, we could practically abolish that complaint; but, be it noted, it would have the effect of preventing the birth of as many normal as of abnormal persons.

This figure represents the simplest possible type of inheritance. It applies to a fair number of other conditions. Some of them, such as night-blindness, diabetes insipidus, and brachydactyly (short fingers), are more or less troublesome, but neither dangerous to life nor disabling. Until the rest of humanity has been greatly improved there need be no serious objection to the breeding of such persons. Others are more serious. For example, cleidocranial dysostosis is an abnormality of the skeleton involving invalidism of a

greater or less degree. Neurofibromatosis is a
disease characterized by skin tumours, which
frequently become cancerous. These, and a
number of similar diseases rarely, if ever, skip
a generation. It would appear that sterilization
of the unfit would abolish them.

However, recent work has shown that in the
more dangerous of these diseases such a hope is
vain. They would disappear as the result of
natural selection were it not that sporadic new
cases constantly appear as the result of mutation.
Hence sterilization (or other measures to prevent
the breeding of affected persons) would abolish
a majority, but not all, of the cases. At any rate,
no eugenic measure, except the prevention of
breeding by unfit persons, is called for.

Unfortunately, things are not always so simple.
Let us take the case of hereditary brittleness of
the bones, such as occurs in some families. In a
severe case dozens of fractures may occur, some
even before birth, and the sufferers are hopeless
cripples. Now in these families all the brittle-
boned members have blue sclerotics, that is to
say, what in normal people are the whites of the
eyes are of a dull bluish-grey colour. Vision is
normal, but deafness is common. Blue sclerotics
are due to an autosomal dominant gene, and are
handed down to half the progeny, without
skipping a generation.

Fig. 2 is the pedigree of a family in which that
defect occurs. The black symbols there represent

persons with blue sclerotics. The letters F and D refer to their more serious abnormalities. It will be seen that some of the persons with blue sclerotics had multiple fractures and others were deaf, while two were affected in both ways. But a very

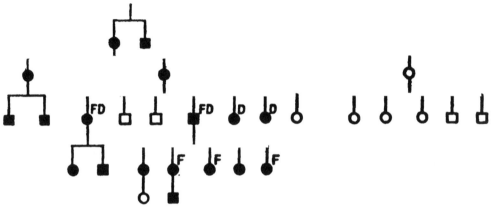

FIG. 2.—Pedigree of blue sclerotics and associated diseases (all spouses normal).

□■ Males, ○● Females.
□○ White sclerotics, ■● Blue sclerotics.
F Fractures, D Deaf.
(From the *Treasury of Human Inheritance.*)

considerable number of persons with blue sclerotics escaped from either of those complaints, although in their children the blue sclerotics might be associated with bone fragility or deafness. If we were to control breeding so as to wipe out those conditions, we should have not only to prevent the breeding of the deaf people and the people with brittle bones, but we should have to say that anyone in that family with blue sclerotics ought to be sterilized because their children might have brittle bones. In other words,

we should have to extend our principle of
sterilization not only to the unfit, but to the fit.
Indeed, in a disease of this type sterilization of
the fit is much more important than that of the
unfit, because it is fairly obvious that people
with multiple fractures due to brittle bones are
not likely to breed to any very great extent. If
we were to apply sterilization in this case, we
should prevent the birth of one unhealthy child
at the sacrifice of two or more healthy ones.

A good number of diseases are inherited in
the same way. Among them is pre-senile cataract;
that is to say, cataract affecting people in their
youth or middle age. In most of the hereditary
cases the descent is through affected persons,
but if you study the pedigrees you find that many
of the affected persons marry before the disease
has manifested itself. Several other eye diseases,
such as hereditary glaucoma, show the same
phenomenon. Huntingdon's chorea is a very
terrible affliction of the nervous system, leading
to involuntary movements, and often ending up
in madness, which is transmitted, as far as we
know, only through affected persons. But the
mean age of incidence of that disease is thirty-
five years, and sometimes it does not appear
until sixty or seventy, although it may develop
in early youth. Therefore, a great many of the
sufferers have married and produced children
before the disease has manifested itself. In the
present state of our knowledge the only method

of wiping out that disease would be to prevent the breeding not merely of affected persons but of their children, because they might very well develop the disease after marriage, and after having had children themselves. It is clear, then that the apparently simple principle of sterilization of the unfit, like so many simple principles,

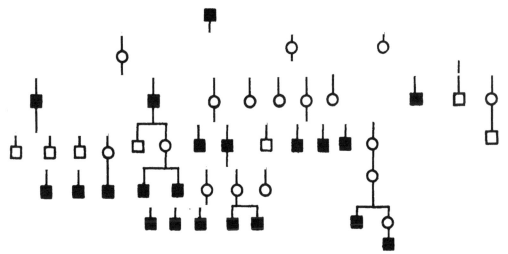

FIG. 3.—Pedigree of mild haemophilia (all spouses normal).
□ Normal male, ■ Haemophilic male, ○ Normal female.
(From the *Treasury of Human Inheritance*.)

if pushed to its logical conclusion, carries you rather further than appears at first sight.

Now let us take a different type of inheritance, the so-called sex-linked inheritance. Fig. 3 is a pedigree of haemophilia, a disease practically confined to males, in which the blood does not clot normally, and which may lead to very serious consequences. There are two forms of it, the mild form shown in Fig. 3 and the severe form shown in Fig. 4. The principles of inheritance are the same in both. It is never, or very rarely, handed down by an affected male to his sons or to his daughters. Nevertheless, the daughters, though not the sons, hand it down to approximately one-half of their sons. It will be seen that in one case in Fig. 3 it has been transmitted

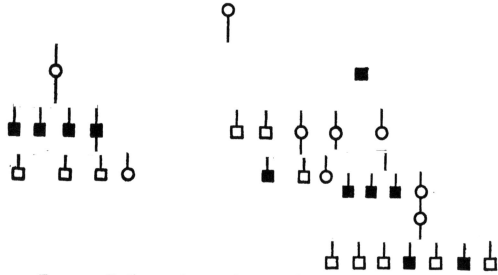

FIG. 4.—Pedigree of severe haemophilia (all spouses normal).
□ Normal male, ■ Haemophilic male, ○ Normal female.
The woman at the head of the pedigree had seven other children, sex and condition not recorded. Seven of the eleven haemophilics died of the disease. (From the *Treasury of Human Inheritance*.)

from the male at the top through no less than five generations of women before it finally affected the male at the right-hand bottom corner. The more usual type of pedigree is that of Fig. 4, in which the vast majority of males do not live long enough to have any children. Preventing the males from breeding would have done very little good for the simple reason that the majority of them died in childhood, and therefore were cut off before they could have any children. One may add that in this particular case the operation of sterilization of an affected male would probably lead to death through haemorrhage.

The question then arises whether the women who are carriers should be sterilized. There is a case for it. But supposing we were to sterilize all mothers and daughters of haemophilics we should prevent the coming into the world of three perfectly healthy children for every unhealthy one. If we extended it to their sisters we should prevent the birth, on the average, of seven healthy children for every unhealthy one. I do not know where we are to stop in a case like that. One would certainly stop at the sacrifice of a hundred healthy children to prevent the birth of one diseased one. I want to point out, though, that it is an extremely difficult problem. It is not one which can be decided on any simple principles.

Further, no amount of sterilization would wipe out haemophilia altogether. The deaths of haemophilics are unfortunately compensated for by the sporadic appearance of the gene through mutation, so that the frequency of the disease remains fairly constant. Probably at least a quarter of all cases are sporadic, and could not be avoided by any eugenic measures at present available. This does not, of course, mean that eugenic measures would be useless.

A fair number of other diseases are inherited in the same manner as haemophilia, some, like colour-blindness, being trivial, others of varying severity. In some cases the disease is mainly confined to males, but a small proportion of women, corresponding to a fraction of the women transmitters of Figs. 3 and 4, manifest the disease. In this class are hereditary optic atrophy, one of the causes of blindness, and anidrosis, in which the sweat glands are absent and the teeth defective.

In all the cases so far considered there is a sense at least in which one can speak of the affection as hereditary. It is handed down sometimes from parent to offspring, sometimes from grandfather to grandson, skipping a generation, and so on. We now come to a group of diseases which are congenitally determined, but are not hereditary in the ordinary sense.

These abnormalities are due to what are called autosomal recessive genes. A person carrying one such gene is entirely normal, but a person carrying two is abnormal. In most cases the abnormal gene seems to be a more or less completely inactive form of the normal one. In a person with two abnormal genes some important function is not performed at all, or is performed very inefficiently. On the other hand one normal gene will do the work of two, just as (to use a possibly slightly misleading analogy) one kidney can do the work of two, so that a man may go through life with a single kidney and never know it.

The majority (or in some cases all) of the abnormal persons arise from the union of two parents, each of whom carries a concealed normal gene—heterozygotes, as they are called. Fig. 5 is a pedigree of juvenile amaurotic idiocy. The victims of this disease are born normal. At the age of about six they gradually go blind. They

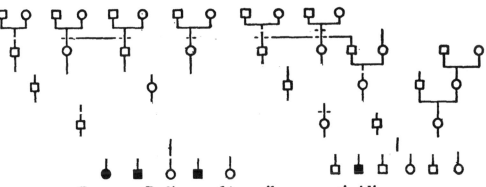

FIG. 5.—Pedigree of juvenile amaurotic idiocy.
□■ Males, ○● Females.
□○ Normal, ■● Idiots.
(After Sjögren.)

then become progressively idiotic. They are generally hopeless idiots by fourteen, and dead before they are twenty.

In Fig. 5 it has been necessary to show both spouses in the marriages because both may be concerned in transmission. The fathers of the idiots were brothers, and each married a cousin. The recessive gene was handed down to the parents of the idiots from one of the two spouses shown at the left-hand top corner.

This pedigree is taken from the work of Sjögren, who went through all the schools for blind children in Sweden, and probably detected almost all the cases in that country, 155 in all. The proportion of children destined to develop the disease is about 38 per million. In the general population the frequency of first-cousin marriages is about 1 per cent. Among the parents of idiots of this kind it is 15 per cent; while at least another 10 per cent are blood relations of some kind. A further calculation shows that roughly one Swede in one hundred is a heterozygote for this gene, that is to say, carries one gene for juvenile amaurotic idiocy, which can only be detected if he or she has children by a similar spouse.

Now, apart from inbreeding, only one heterozygote in a hundred will marry another heterozygote. That is to say, one marriage in ten thousand will yield idiots if enough children are born, the average proportion being one quarter. Thus one in forty thousand or twenty-five per million would develop this form of idiocy if marriages were at random. But if a heterozygote marries a blood relation the chance that the spouse is also a heterozygote is much increased.

Genes of this type cause a good many very serious abnormalities. It is clear that sterilization of the unfit would be entirely ineffective in this case, since there is no record of a victim of this disease ever producing children. Sterilization of heterozygotes is also impossible, both because there are far too many of them for it to be practicable, and because the vast majority of them are undetected.

There are two methods available for dealing with this type of disease. The first is to prevent or discourage the parents of children who have ever produced an abnormal child of that type from having further children, either by allowing divorce between them or otherwise. The second method is to discourage inbreeding. Curiously enough, the only body in this country which does discourage inbreeding is the Catholic Church, which is strongly opposed to many other eugenic measures. It is not the case that by preventing the marriage of first or even second cousins you could wipe out that disease or similar diseases, but you could do a good deal. For example, of the cases of congenital deaf mutism a fraction, varying in different countries, of between 20 and 40 per cent are the children of first cousins. Of the victims of retinitis pigmentosa, a disease which accounts for about 1 or 2 per cent of blindness in this country, about one-third are the children of first cousins. Among the parents of children suffering from a very terrible disease, *xeroderma pigmentosum*, a skin affection which usually ends up in cancer before the age of fifteen, something like half are close blood relations.

As regards physical defects, therefore, two types of eugenic measure are desirable: first, to discourage the breeding of affected persons, and, in some cases, of certain of their relatives, and secondly to discourage inbreeding in general. The relative importance of those two measures depends on the relative frequency in the population of dominant abnormalities; that is to say, abnormalities which are handed down by affected persons to an appreciable proportion of their progeny; and recessive abnormalities, that is to say, abnormalities which show up mainly as a result of inbreeding. Unfortunately, we know very little as to the relative frequency of these two types of abnormalities in man. My colleague, Dr. Gordon, in a population of wild flies, found that something like 50 per cent of the individuals carried a recessive gene of some kind, which, although the flies themselves were perfectly normal, showed up if their progeny were strictly inbred. But we cannot argue with any certainty from that analogy to man. We can only say that, in the majority of animals so far studied, most abnormalities are recessive, that is to say, of the type brought out by inbreeding, rather than dominant, that is to say, of the type directly handed down. It is perfectly possible that man is exceptional in that respect. It is, to my mind, of very considerable importance that we should know whether or not man is an exception before we frame any thoroughgoing eugenic policy.

As regards the dominants and the sex-linked

recessives, that is to say, characters which are handed down to a marked degree by affected persons, six possibilities are open. The first is to discourage their marriage, and the second is to forbid it; the third, to encourage married chastity. The fourth is to encourage birth control by affected persons; the fifth to encourage their voluntary sterilization; and the sixth to encourage their compulsory sterilization.

Now I would suggest there is at least a case to be made out for trying the milder measures before the severer ones, and the voluntary measures before the compulsory ones. In the case of mental defect some is unquestionably due to nurture, to injury at birth, to diseases of various kinds, and to faulty pre-natal environment. Some, again, is due to the recessive genes of the type which I have described. Only a small proportion is due to dominant genes; that is to say, only a small proportion of our registered mental defectives are themselves the children of defectives. For example, in Birmingham 345 children were investigated, one or both of whose parents were certified mental defectives. Of those children only twenty-five, or 7 per cent, were themselves either attending special schools for defectives or under examination for such schools. The others were on the whole below the average in intelligence but by no means grossly so, and how far this may have been due to their unsatisfactory family environment it is difficult to say. Penrose, one of the authorities on the subject, takes the view that about 5 per cent of our mental defectives have one or both parents defective. Others give a somewhat larger percentage. I think it is very difficult to suppose that the sterilization of all mental defectives would lower the percentage in the next generation by as much as 15 per cent. Some writers and speakers on the subject completely neglect the role of inbreeding. There are a few facts which suggest that inbreeding may be of considerable importance. In children in an eastern county investigated by Russell, of thirty awarded free places the parents of two were born in the same village. Of sixty-three with an intelligence quotient of less than eighty, that is to say, definitely backward children, the parents of twenty-five were born in the same village. If that is a general rule, I think it suggests that the rural motor-omnibus service may be quite an important eugenic agency.

There is very general agreement that mental defectives are not likely to make good parents, whether or not they hand on their condition by biological processes to their children. But I am strongly of opinion that, in their own interests, mental defectives should be segregated. It is said that mental defectives are often capable of earning a living, and of playing a useful part of some kind in society. I take the view that any person who, in the present conditions of unemployment, is able to keep a job and be of use to society, should not be rated as a mental defective, even though he may not be quite as intelligent as some of ourselves. The sterilization of mental defectives as an alternative to their segregation has been suggested very largely as a measure of economy. How far any given economy is necessary is not for a biologist to state; but a biologist may, I think, point out that the demand for the sterilization of such persons is primarily made on economic and not on biological grounds.

It is worth while examining the origin of the belief that sterilization or similar methods would go far to solve social problems. First, I think, comes the false analogy with domestic animals. In domestic animals we select in the most rigid manner for desirable characters by castrating or killing a large majority of males, by only breeding from selected females, and, above all, by fixing such characters as we have got by fairly close inbreeding. There is no question that by inbreeding the large majority of characters in animals can be fixed so that they are manifested in all, or nearly all, of the progeny. But, in the course of inbreeding, a large number of abnormal types appear in the early generations. In the case of animals they are destroyed. I think a policy of that kind, a eugenic policy involving inbreeding, might be quite suited to a society in which defective children were killed off. But rightly, as I think, our society takes a different view as to human life, and unless we are to alter our ethical code radically we have to avoid inbreeding. We must realize that we cannot fix those characters which we regard as desirable. Therefore, although we could undoubtedly get rid of a certain number of defectives by sterilization, or by less drastic measures, we cannot hope to, and perhaps it is a very good thing that we cannot hope to, attain in man anything like the degree of fixity which is possible in domestic animals. Secondly, the belief that sterilization was somewhat of a panacea found, as I believe, wide and uncritical acceptance as a result of the class struggle. A section, though in this country only a minority, of the well-to-do listened perhaps rather too eagerly to arguments against public health measures and humane treatment of defectives in institutions because they were both expensive, as they undoubtedly are.

Thirdly, I think the desire to dominate over human beings which finds an outlet in war and in despotic forms of government has manifested itself to some extent in this campaign for sterilization. I do not think myself that it is accidental that the National Socialist movement in Germany, which has dissociated itself from so many ideals which are regarded as civilized and Christian in other countries, has also associated itself with the programme of wholesale sterilization.

There is to-day before the country a so-called Voluntary Sterilization Bill. You will find its text in the *Eugenics Review* for July 1935. It permits of the sterilization of four classes of persons. First, the mentally defective; secondly, those who have suffered from mental disorder in the past; thirdly, those who have a grave physical disability which is deemed liable to be inherited; and, fourthly, persons who are deemed likely to transmit a mental defect or a grave physical disability to subsequent generations. You will notice how very wide that last clause goes. Almost any female relative of a haemophilic on the mother's side has a certain likelihood of transmitting that disease. It may be that there is a chance of one in eight, or one in sixteen, or one in thirty-two; but there is nothing in that clause to say where one is to stop, and that in an Act of Parliament is rather an important defect.

For sterilization a medical certificate from two doctors would be required, one of them to be approved for that purpose by the Ministry of Health. The consent of the person concerned would also be needed if he or she were not mentally defective; in the case of a mental defective either the consent of the parent or guardian would be required, or, if they were not to be found, that of the local authority responsible for the defective. Further, in the case of persons under twenty-one, other than mental defectives, the consent of the parent or guardian would in any case be necessary. It is interesting to note that the consent of the husband or wife of the person to be sterilized is apparently not necessary, although they have to be informed of it. Now I would go so far as to say that in England at present there are at least six medical men and one medical woman who are sufficiently educated in genetics to allow them to act as thoroughly reliable referees in such cases; but it is important to realize that before such an Act is workable systematic instruction in genetics in our hospitals is absolutely necessary. There is no such systematic instruction at the present moment.

The title of the bill is somewhat misleading because, of course, in the case of mental defectives sterilization would not necessarily be voluntary. They might not understand what was happening to them. What is more, it is fairly easy to persuade a mental defective that any given course is desirable.

I think the main objection is the following. The operation of sterilization, which is not, of course, castration, is a trivial operation for a man; but for a woman it is about as severe an operation as, for example, an operation for uncomplicated appendicitis. It would very rarely be fatal; nevertheless, in a sufficiently large number of cases, you would have an occasional death.[1] Now it is a general principle of English law that we must not endanger the life of others, even with their consent, much less without their

[1] Since this lecture was delivered, a woman has died in Denmark as a consequence of sterilization.

consent, except in order to avert some greater danger to them. If we kill such a person, for example, by procuring abortion, even although it is done with their consent, we are guilty of murder, or, at best, manslaughter. That principle, which we may call the sanctity of human life as applied to medicine, is a very broad principle, and it may be desirable to change it; but I do submit that to throw it aside would be something much more revolutionary than, let us say, nationalizing the Bank of England. It is not a principle which should be rejected without very thorough consultation of the electorate. If a majority of the people, after hearing both sides, decide that in the interests of subsequent genera-tions, or in the interests of economy, it is desirable to endanger the lives of certain women, we may have to put up with it; but do not let us have this bill passed as a non-contentious measure, and, above all, do not let it be pretended that all, or even the majority, of scientific workers in the field of genetics, though they may support voluntary or perhaps compulsory sterilization of men, hold the same view as regards women. Personally, I may say that I see no objection to the voluntary sterilization of normal males with a dominant physical defect, that is to say, one which will be transmitted to about half of their offspring, provided that such sterilization could be made a ground of dissolution or nullity of marriage if the wife desires. But I do think that

the bill which I have summarized would lead to
considerable injustice, as similar laws have in
the United States, and very possibly to a violent
reaction not only against that particular application of human genetics, but against other
applications of it, such as the discouragement
of parenthood by certain persons, which I, at
least, believe to be eminently desirable, and
which I think that most people who have seen
examples of certain inherited abnormalities would
agree was to be wished.

Now I want to touch briefly on the race
question. Our difficulties there are largely difficulties of definition. We say that an Englishman
is of a different race from an African negro. We
also say that he is of a different race from an
Italian. In the first case, that of the negro, as
regards certain characteristics there is no overlap
at all between the two races; that is to say, the
blackest Englishman is lighter than the whitest
negro, and so for a number of other physical
characters. If you are shown two people you can
be perfectly certain which is the Englishman
and which is the negro. But as regards the Italian
there is no such certainty. It is true that most
Italians are darker than most Englishmen, but
it is very easy to find fair-haired Italians and
dark-haired Englishmen. An attempt has been
made to divide up the population of Europe
into a number of races, of which the Nordic,
Alpine, and Mediterranean make up most of

the population of Western Europe. The Nordics are tall and long-headed, with fair hair and blue eyes. The Alpines are short and round-headed, with brown hair and rather variable, but usually brown eyes. The Mediterraneans are short and long-headed, with black hair and dark eyes. It is perfectly true that you can pick out types of this kind, and that you find the Nordic type much more frequently in Sweden than in Italy. But if we take any of the physical characters by which these races are differentiated, for example, head length, we find an enormous overlap. Consider the Swedes, one of the longest-headed peoples in Europe, and the Bavarians, one of the shortest-headed. If we make graphs of the distribution of their cranial indices, that is to say, ratios of head breadth to length, we find that we can only speak of averages. If you take one hundred Swedes and one hundred Bavarians at random, and measure their heads, you can be quite sure which group is which. But if you take one Swede and one Bavarian, and try to identify them on the basis of their head shape, you will be wrong in about 14 per cent of cases. The two populations overlap.

This overlap is characteristic of any measure you like to take to differentiate these so-called races in Europe. They are at most statistical conceptions.

So much for our definition. We may mean two quite different things by races: first of all, groups which do not overlap, and, secondly, groups which do overlap. I prefer the former meaning.

When we say that race A is superior to race B in some particular respect, let us say musical ability or moral behaviour, we may mean several different things. First, we may mean that there is a difference, but it is largely due to their environment. For example, one race was converted to Christianity as our ancestors were fifteen hundred years ago, and the other has not been so converted. Again, in the pre-Columbian races of America, such as the Aztecs, we find evidence of great cruelty, and of human beings treated as animals; and no wonder. In pre-Columbian America there were practically no domesticable animals—the llama was the best—therefore in agriculture men had to be used for performing mechanical tasks which in the early stages of our own civilization were performed by cattle and horses. You cannot wonder that they were regarded as our ancestors regarded cattle and horses, something to be slaughtered.

If, however, we are referring to innate or biological superiority we may mean one of four things. First of all, we may mean that there is no overlap between the races, and that the worst of race A is better than the best of race B. There is quite obviously no case of this kind in reality. Some coloured races may perhaps be on the average not quite so clever as ourselves, but some of them are certainly cleverer than the worst idiot in London.

Secondly, we may mean that race B has some upper limits which race A has not. It has been stated that the brains of Bushmen are of a more primitive type than those of other human beings, and that for that reason (though to my mind it does not follow) Bushmen are incapable of rising to certain levels which you and I attain. Even if that is so it is rather unimportant, because the number of Bushmen in the world is very much less than the number of mental defectives in London. It is also alleged that an abnormal cell arrangement is found in the brains of East African negroes. As it is not observed in West Africans, who differ still more from the European type in other respects, it may very possibly not be due to inheritance at all, but to food deficiency, which is extremely common in East Africa. If it is an inheritable character it still has to be shown whether it is any real handicap to mental development.

I think we may say that there is no clear case in which race B has a definite upper limit set to it.

A third possibility is that the average of race A is better than the average of race B. More accurately, we should say the median, because you cannot average a quality like ability. If you take a thousand and one people and put them in a row, the ablest on the right and the stupidest on the left, and if you then take out the middle one, he gives you a very good representation of the average, although there is no such thing as average ability in the sense that there is average weight. That middle person is called the median. There is no sort of evidence for differences of this kind between the modern European races. In America, as you will remember, there was at one time a belief that the Nordics were superior as immigrants to the Alpines and the Mediterraneans, so they got a Jew, who might be regarded as impartial, to examine children in a number of European cities and rural areas. The differences were extremely small. However, the Parisian children, who were predominantly Alpine, turned out to be a little more intelligent, according to his tests, than those of the Nordic Hamburg or Mediterranean Rome. On the other hand, the Mediterranean "race" came well to the fore in producing the most intelligent group of country children in the French Pyrenees. The stupidest lot but one were "Nordics" from French Flanders. All that can be said from such data is that whatever else may be correlated with the physical differences which are ascribed to the Nordics, Mediterraneans, and so on, intelligence of the kind tested by these tests has a very slight relation to it.

On the other hand, when it came to the white and coloured populations in the United States it was found that in the intelligence tests for entrance to the army during the War the whites did very much better than the negroes; and it was held that those intelligence tests revealed innate qualities to a certain extent. That may be true, but some rather remarkable facts point to a different conclusion. In five northern states the average scores of the whites was 68 and of the negroes 40; but in eight southern States the whites averaged 40 and the negroes 19. In other words, the northern negroes secured as high marks on those tests as the southern whites. Indeed, the negroes of Ohio were ten points better than the whites of Arkansas. Granted that, I think we must take it that environment as well as race plays an important part in this difference in intelligence, and we certainly cannot be sure, even though we may suspect, that race as well as environment accounted for some of this difference.

The fourth possibility is that there may be more exceptional men and women in race A. For example, when we say that the Germans are a musical people we think of great musicians like Bach, Beethoven, and Wagner, who occur with a frequency of about one in fifty million, or even more rarely. It may be that superiority of that kind merely means that the people in question is more varied in their innate endowment; it does not necessarily mean that their average is better. Now I think it highly probable that innate differences of the last two kinds, namely, in the average and in variability, do occur, but I would point out that there is very little, if any, scientific evidence for them. Nowadays the belief is widely held in Germany—I quote from Professor Günther—"that a race is a group of men which is differentiated from every other human group by its characteristic combination of bodily characters and psychological peculiarities." That is quite certainly often false as regards the distinction between different human groups, even for physical characters. It is true as regards some physical characters for the distinction between Europeans and negroes, but I have yet to learn of any psychological characteristic which is found in all negroes and in no Europeans. There may be such, but it has yet to be discovered.

This fallacy has, I think, arisen from two false analogies: first, with species of animals which do not generally breed together, and differ very much more than human races; and, secondly, with races of animals, for example, greyhounds and bloodhounds, which do not overlap in their behaviour. The one hunts by sight and the other by smell, and this is correlated with physical differences. Such characters, however, have been fixed by inbreeding and selection, and there are nothing like such pure lines in men. Nevertheless, it would be theoretically possible to produce such groups. Occasionally a pair of boys or girls is born with exactly the same assortment of genes. They are called monozygotic twins. They show a remarkable resemblance, not only in their physical traits, but in their character and intelligence. Thus, if one member of a pair is a criminal, the other is generally a criminal of the same type, although this is not true of ordinary twin pairs, even though they are brought up in the same environment. It would be possible by inbreeding to fix characters in human beings so that in twenty generations or so one group had a strong tendency to crime or virtue of a particular kind, but this could only be done by such intense selection and in-breeding as has been practised in greyhounds. It has never been done in human beings, with the possible exception of small inbred groups, such as the Incas in Peru, and, to my mind, it is very undesirable that it should ever be done, if only because we do not know enough about human ideals to say what type we should breed for, although we do know enough to say that certain grave defects are undesirable.

The recognition that there is as yet no scientific basis for current dogmatism about racial differences does not support either of two extreme theses which are, so to speak, the opposite of the views held in Germany. The first of those theses is that migration between different countries should be perfectly free. I take the view that the Englishman is on the whole superior to the negro in England, but I also take the view that the negro is in many ways superior to the Englishman in West Africa, which is commonly known as the white man's grave. I therefore do not think that it is biologically or socially desirable that there should be free migration between England and West Africa, although I would not prohibit any migration. It seems to me that it will generally be bad for the migrants in each case.

The second thesis which I cannot support is the thesis that racial crossing is desirable. Here the evidence is extremely inadequate, but it is worth noticing that in cases of crosses between different animal varieties the first cross is generally very vigorous, and little or no harm occurs if it is repeatedly crossed into one or other race. On the other hand, if hybrids are bred together the progeny are often less vigorous than either of the original races, or than the first generation of hybrids. There is some reason to think that in South Africa, for example, the so-called Cape coloured people, who are derived from a mixture of races and who now breed together, are less healthy than either the whites or the Bantu negroes. It does not follow that the first cross is harmful in such a case, and it may be that in the case of some human races no harm whatever comes either in the first or in the subsequent generations from the mixture. But it is perfectly possible that, from the cross of two races which differ, though each has its own admirable quali-ties, you may in subsequent generations, though probably not in the first, get something worse than either. I do not know. I do not think we ought to forbid racial crossing. I do not think we ought to encourage it either, and I do think that we ought to study it.

The result of our investigations has been largely negative, but I think they have shown that certain eugenic measures are desirable. On the other hand, we have sound reason to reject the more extreme views based on a false analogy with domestic animals, views which, as I think, are definitely not in consonance with the ideals which our Chairman has founded this series of lectures to investigate. But, on the ground which we have cleared this evening, I hope that in this and future generations we may be able to erect a scientific policy founded on the facts of human genetics.

VI

PROFESSOR JULIAN HUXLEY

CHAPTER VI

SCIENCE AND ITS RELATION TO
SOCIAL NEEDS

WE have been repeatedly told that ours is a scientific age; but it is also an age which is in rather a bad way. Trying to relate this fact of the prevalence and development of science with the economic and social troubles of the times, one is forced back to fundamentals. One ought to begin, I think, by asking what really is science, and how is it related to society?

There are some who find in science the evil genius of to-day, and who would like to see scientific research given a holiday, or even to see a return to a more medieval state of affairs. But when one says "science," what does one mean? Is science just pure knowledge, or is it a means to practical ends, a sort of mechanized Santa Claus? Is it advanced by man's disinterested curiosity, or is it merely the paid servant of business or politics? Is it dispassionate, or consciously or unconsciously biased? When it makes its advances, does it advance along its own road, or is its road determined for it by the economic system and temper of the time?

I think that as a matter of fact all these ideas are in their degree true. There have been many who have preferred the laborious pursuit of

knowledge to pleasure, fame, or wealth. One
has only to think of Leonardo da Vinci, Henry
Cavendish, or Charles Darwin. There is such a
thing as disinterested curiosity, and it may lead
to scientific discovery. But scientific discovery
can also be applied. There is a well-known story
of Faraday in his laboratory being visited by
some distinguished statesman. I do not remember
who it was—shall we say Mr. Gladstone or Mr.
Disraeli?: you can choose according to your
politics. On leaving, the great man said, "Mr.
Faraday, this is very interesting, but will you
kindly tell me what is the *use* of all your work?"
And Faraday replied, "Perhaps you will allow
me to ask you a question in return: Could you
tell me what is the use of a baby?" With all
deference to mothers, this was a difficult question to answer. One cannot tell—until the baby
grows up. Faraday's baby, as we all know, grew
up into the major part of the modern electrical
industry.

Another striking example is that of Pasteur,
setting out to discover whether life is spontane-
ously generated or not, and becoming the father
of modern surgery and the germ-theory of infectious disease.

This is inevitable when you think of it. Thought
is in a sense action short-circuited, and knowledge
must hold the potentialities of control, so that
every well-founded scientific theory cannot help
containing an infinity of applications.

This idea that in the development of science, pure knowledge generally comes first and leads later to practical application is what you generally find stated in the textbooks. On the other hand, the reverse is also true. We all know that action may help knowledge; practice may show up the defects of theory; and tangible needs may stimulate one to think and inquire; accordingly we find numerous examples of the way in which the current has flowed from practice to theory rather than in the reverse direction.

For instance, steel practice and needs have given us new and fundamental knowledge of the behaviour of alloys. Or, again, recent work, of a purely practical nature in its inception, on the application of X-ray analysis to the fibres of wool, is giving us fundamental new knowledge of the nature of the protein molecule, and is so shedding light on some of the major problems of biology.

Thus the current may flow in either direction. Either knowledge may lead to practical control, *or* practical needs may stimulate the spread of pure knowledge. However, neither of these things may happen. In the Middle Ages there was little scientific curiosity, and what there was was discouraged; in ancient Greece there was little practical control compared with the amount of theoretical knowledge. We see that scientific curiosity may be stifled or turned into other channels, and that the usual connection between practice and theory may not hold. It was the theological system in the Middle Ages which was responsible, and in the case of ancient Greece it was the economic fact that the industrial application of science was not necessary because the economic system was based on slavery.

What happens to-day? After all, the scientist (unless he be a rich man, and it is rare for scientists to be rich) can only work if paid, and he can only be paid in the long run because somebody, whether industry, the State, or the private employer, thinks it is worth while paying him to do it. Naturally, the amount of money spent will determine the emphasis on different branches of science. Take the single rather striking example of that great industrial firm, Imperial Chemical Industries, which supports a number of research laboratories. In one of these alone there are more trained research chemists working on chemical problems than there are trained psychologists and social scientists doing research in their sciences in the whole of the kingdom. Obviously chemistry cannot, in such circumstances, help advancing faster than social science and psychology. Or take another example: perhaps four or five times as much money is being expended in this country on industrial research than on research promoted to improving human health.

I think that now we are beginning to see something of what science really is. It is neither something abstract, not just a product of a few inventive geniuses, nor, on the other hand, is it a mere mechanic jade or the servant of profit-making or State interests, but it is a resultant of a number of forces: first of all, the natural disinterested curiosity of certain men; secondly, the desire of others to get things done and acquire control over nature; thirdly, the general temper of the time; fourthly, the amount of money spent by those who control the purse-strings; then the nature of the general economic system; and finally (a point I have no time to go into), the state of technical efficiency of the period, which will often determine what scientific work can be done, and what cannot. Just like any other human characteristic, science, too, is the product of an interaction between something hereditary and the environment in which it has to blossom.

So science is a social function, a particular type of social function, just as circulation or digestion is a function of the body. As a social function, is science to-day doing its job as efficiently as it might? We all know that in different organisms their functions are at different levels of efficiency—for instance, intelligence in an earthworm is quite definitely inferior to intelligence in a monkey.

Obviously, science is doing a great deal. I need not emphasize that. It is this aspect of science which forms the stock-in-trade of articles on popular science. One need only think of science and steel, the hydrogenation of petrol, the cure of diabetes, anaesthetics, plant-breeding, and so on.

It is fair to say that science and its results are the basis of our civilization, just as the steel girders in a modern building are the basis of the building. Take away the results of science and our civilization would collapse as the building would collapse without its girders; and further, without science there would be no hope for the advance of civilization in the immediate future, let alone the remote future.

But one sees, when one goes into the question more deeply and does not merely take the superficial optimistic view, that there is a great deal of mal-function. Science is often frustrated, or diverted, or distorted, or not applied, or not done properly, or not done at all. Let us take a few examples of this view, which is not so often stressed as its opposite.

The most obvious case of science not being applied is in agriculture. You get the scientist, as a result of Mendel's epoch-making theoretical discoveries, synthesizing new types of wheat, just as the chemist would synthesize new compounds in a laboratory, thus increasing the yield per acre, and making it possible to grow wheat nearer the Pole and farther into the desert regions of the world. And yet the practical outcome is that the great wheat-growing countries of the world are now all trying to reduce their

production and cut down the acreage. Science
here is very far from being applied!

Another case concerns the extension of good
grassland over barren moors, as made possible
by the work of Stapledon at Aberystwyth. There
is no chance of this being applied on the large
scale actually possible, which might add new
pastures equal to one-sixth of our total area, and
double our population of sheep, in present
economic conditions.

Then take another example, the application of
science to building. All sorts of things are being
done by science to improve building and make
it more efficient, healthy, and comfortable; and
yet these efforts are often frustrated by old
specifications and by-laws, and for economic
reasons. We may especially note the lack of
incentive to mass-production in housing, which
might be as revolutionary as mass-production in,
say, motor-car manufacture, or anything else.
We could, with our scientific knowledge, re-house
the whole population efficiently, healthily, and
beautifully. We are now making a beginning
with it, but our total effort is belated and inadequate.

Or take smoke: why do we still allow anybody
to burn raw coal, either in a domestic grate or
in a factory? A hundred years ago that was the
natural thing to do, but to-day scientific knowledge not only permits us to say that it is extremely
wasteful, and causes material loss, and damages

health, but also has shown us better ways of utilizing the resources of coal—turning it into electricity, or giving us gas and smokeless fuel and valuable chemical products, or hydrogenating it to petrol. We do not prohibit the burning of raw coal. Why?—for so-called economic reasons, because smokeless fuel is still slightly more expensive than normal fuel. This is a good example of one reason why science is not fully applied —namely, the lack of a body capable of taking a really all-round view. To prohibit the burning of raw coal would not pay in the short view; but supposing we could establish an all-round balance sheet with all departments of national life represented, in which damage to health and agriculture and buildings and so on could all be totted up and put against the saving effected to the actual users of coal in £ s. d., I venture to think that we should find it *would* pay, hand over fist. There we have an example of how necessary it is to have some co-ordinating mechanism concerned with the application of science.

We may trace an analogy in regard to biological evolution. In the course of biological progress you get the mind and brain developing steadily. But you find extremely efficient sense-organs and elaborate and efficient instincts developing long before you reach the crowning gift of evolution— the capacity of the higher organisms and ourselves of co-ordinating those avenues of knowledge and action so that they may be adjusted to distant goals and varying circumstances. We have not yet reached that stage in the evolution of our societies, but to do so is urgent.

Perhaps an even more striking example of unapplied scientific knowledge concerns health. There we can see just how in very recent years knowledge has increased, and what it might do. I refer to the growth of our knowledge of the vitamins and other accessory food factors during the last twenty-five years. As a result of this work, it is safe to say that the general level of physique, of alertness mental and physical, and of disease-resistance in our own population and that of all civilized industrial countries could be enormously raised. If we had a benevolent dictator (always granted dictators can be benevolent!), then just by telling the people what food they should eat, and providing the relatively small amount of money necessary to provide the extra cost, he could add an inch or probably two inches to the average stature and six, eight, or ten pounds to the average weight, enormously reduce the incidence of infectious disease, and increase general alertness. Why has this knowledge not been fully applied? Partly because the knowledge is so new—there is always a time-lag in these matters. Partly because you have not got a co-ordinating body to think of the different aspects of these modern researches; and partly again because there is no economic incentive.

Of this last point we have a very illuminating

illustration in the shape of what is undoubtedly a very important step forward in the right direction, the provision of free milk to underfed children. That is the thin edge of the wedge of a rational and national food policy. But why was that step undertaken? It was undertaken in the first place not at the instance of the Ministry of Health, nor at the instance of the Board of Education. It was undertaken at the instance of the Milk Marketing Board, anxious to find an outlet for its surplus milk. You may laugh, but that, I understand, is sober fact.

This illustrates a very interesting and important point, namely, that both scientific research and its application is far more efficiently organized from the standpoint of the producer than that of the consumer, far more so from the standpoint of the State than from the standpoint of the individual citizen. And the reason is fairly obvious—because there is a profit incentive in the one favoured case, and a mixture of profit and survival incentive in the other. The consumer and individual citizen are relatively unorganized, and their needs are usually needs which cannot be assessed in terms of hard cash.

Then you come to fields where scientific research is hardly or not at all carried out, at any rate by organized State-aided bodies. Take the subject of population research. You would think that this was a matter of vital interest to all Departments of State, and yet we find the only careful survey of the future trend of population in this country has been done privately, by academic scientists working in universities. Or take the subject of birth-control. You may agree or disagree with birth-control, but nobody can deny that it is one of the major factors which have come into the world in recent years; yet all the research of a fundamental character being done on it is being carried out by small, privately financed bodies. The same is true of research on eugenics—the long-range problem of how to improve the racial constitution of the nation.

We may now consider scientific research in the aggregate. Take industry, war, the different branches of the fighting services, agriculture, and medicine. They all have their scientific research councils, run under the State. In the field of economics there is an economic advisory committee, but not a research council; and there is neither a central research body nor an advisory body, either for psychology or social science, or for that particular branch of the latter in which such work is so very urgent, namely, education.

As a result you find that the structure of scientific research in this country is extremely lop-sided. If you take an estimate of the amount of money spent on the different branches of research you would find that about half or more of the total is devoted to industrial research and pure research in physics and chemistry, which underlie that; next, about half of this amount would be spent on research on war; a good way below that comes research in agriculture and the branches of biology underlying that; still further down comes research on medicine and health and the underlying science of physiology; and finally come social science and psychology, with an infinitesimal amount of money spent on them: as they say in the racing papers, "they also ran."

This lop-sidedness cannot be right. These various branches are more or less co-equal as pure scientific subjects. Indeed, when you come to reflect, far from being right this lop-sidedness is precisely the reverse of the ideal state of affairs.

I remember once hearing Dr. Tizard, the Rector of the Imperial College of Science, deploring the relative lack of money devoted to biological research as compared with physico-chemical. He said that supposing the average man in the street a hundred years ago had been asked what he wanted from physico-chemical science, he would have answered with a list of such things as flying, travelling under water, talking at a distance, rapid communication and transport, the increased use of power, storing up pictures and sounds, and so forth. And all those have been achieved to-day as a result of the application of physico-chemical science in the last hundred years. If, on the other hand, he had been asked what he wanted from biological science, he would have said, presumably, that he wanted more control over domestic animals and plants and pests, that he wanted good health and happiness, increased length of life, control of the development of his children, perhaps the control of sex, and the control of the qualities of the race—something like that.

Now, with the exception of a reasonable measure of control over domestic animals and plants, and a certain, but relatively small, improvement in length of life and health, none of these have been achieved, and in many cases we do not know at all how to proceed to find the answer even in principle; we have not made the fundamental discoveries. Yet we spend perhaps four or five times as much on the physico-chemical sciences and their applications as on the biological.

Even more disproportionate do matters become when we think not of physico-chemical against the biological, but of those sciences dealing with external nature as against those dealing with man himself, with human nature and with those social and economic systems which man has created and which come to have a sort of independent momentum of their own. We all to-day must acknowledge that the world is in rather a bad way. This is not due to any great extent to our lack of control over inorganic or organic nature. We have reasonable control over inorganic nature, but we have not got control over human nature, and we have not got control over the economic, the social and the class systems that have grown up in our midst. We have only to look to national and caste passions, to unemployment and so-called over-production. It is in these fields that control is needed. For this you want first of all knowledge, and yet it is precisely in these subjects that we spend a minimum amount of money and energy on research.

*   *   *   *   *

It will perhaps be helpful to take a few examples to illustrate my points.

Let us take an example of apparently pure science. If you compare a very early telescope such as that used by Galileo with a modern telescope, Galileo's instrument looks insignificant. These instruments have increased our knowledge of nature enormously, and you would think that that was entirely pure knowledge. They give you knowledge of distant parts of the universe, such as spiral nebulae. We now know that each spiral nebula is a universe of stars comparable to our own stellar universe.

Can such knowledge be applied? The answer is—yes, it may have its practical application. Matter as constituted in the remote depths of the universe may reveal behaviour which is of value to the terrestrial chemist or physicist. In addition, such knowledge changes man's ideas and that in itself is a form of control.

The physicists in the Cavendish Laboratory have succeeded in what is popularly called "splitting the atom." That, again, you may say is rather an academic pastime. In a sense it is, but also it is fundamental, and so fundamental that you cannot yet prophesy what practical applications it may have, because they are beyond man's imagination. Already, however, there have been some practical applications; for instance, the discovery of new artificial radio-active substances which may be used for the cure of disease.

Take an illustration from my own science. The great German zoologist Spemann found that if you took a certain little bit of a developing egg of the newt and grafted it into another egg it would give a second set of main organs. The egg is forced to make a second embryo. This you may say is very academic science. But who can state to what that may lead? It may well lead, for instance, to a much greater degree of control than is yet possible over the development of animals and man. In one respect it has already had a very interesting result. This active region is called the organizer. It has been found that it exercises its power by means of a definite substance, chemically related to substances which can artificially produce cancer; and this may be important in cancer research.

Next let us take the view of those who say that science is essentially practical, though it may be necessary to have some academic work behind it. A blast furnace is an example of science as a fairy godmother. It typifies the whole

development of metallurgy in this country. Yet
the achievements of metallurgy cannot be maintained, much less improved, without the application of science. Furthermore, by looking at
different examples of etched steel under a microscope and observing the types of crystals, an
expert is able to tell the practical man what may
be the quality of the steel, and it is through such
scientific work as this that steel methods are
improved.

To emphasize the point let us take another
example of where science seems to be acting
simply as a fairy godmother. A cretinous child
was given thyroid extracts for some time and
completely cured; from being a stunted imbecile
he became normal in physique and intelligence.
Then the parents thought they had enough of
bringing him to the hospital every day; they
ceased to bring him, and the child relapsed into
its original state—until the parents brought him
back for further treatment. And yet such apparently magical results have only become possible
through pure research in the background. Hundreds of anatomists, physiologists, medical men,
chemists, and so forth had to carry out research
on the human body and its chemical and physiological properties before a cretin could be cured.

We now come to the relation between pure
knowledge and its application. It is perfectly
true that science does often proceed from pure
to applied. If you examine the exhibits in that

wonderfully interesting museum at the Royal
Institution of the apparatus of the various great
men who worked there, such as Sir Humphry
Davy, Faraday, and Sir James Dewar, you will
see the pure but modest progenitors of some
important applications.

Dewar worked on gas under low pressures and
temperatures. As a result of that "pure" work
of Dewar's the world has been endowed with
what we now call the thermos flask.

I will come later to Sir Humphry Davy's
work, but now I will refer to Faraday's researches.
The apparatus with which he did his fundamental
work on electro-magnetic induction consists of
rather pathetic-looking coils of wire. Yet one of
the practical results of his work with that
apparatus is the gigantic organization we call
the "grid," by which electrical current is manu-
factured and distributed all over this and other
countries.

Sometimes, however, it is not very clear what
is pure and what is applied, or in which direction
the current is flowing. For example, a spark at a
tension of 1,000,000 volts has been produced
at the National Physical Laboratory. This you
may think is a matter of merely academic interest,
since no one uses a million volts; but in point of
fact such researches are undertaken because the
transmission of electricity at high tension is of
great practical importance—the higher the tension the cheaper it is, and the further you can

send the current. Thus it is no coincidence that in countries where the distance to which electricity has to be conveyed, as in Russia and in the United States, are great, they are operating in their experimental laboratories with even greater voltages—up to 3,000,000 volts.

But in Humphry Davy's safety lamp for mine work you have an example of where the current of science had been flowing in the opposite direction to that usually stated by writers of scientific textbooks. He did not make some theoretical discovery which was then applied in practice. As an eminent man of science he was asked whether he could discover something which could give light in mines and yet avoid the risk of explosion, and with the aid of his scientific knowledge he was able to do so. Here the stimulus to discover was practical need.

The work of the staff of the low-temperature laboratory at Cambridge supplies another example. If you go there you will see men dressed in warm clothes, studying the way apples and other fruit breathe at low temperatures. In that way all sorts of interesting data have been discovered which would not have been discovered otherwise, and pure knowledge of plant physiology has been increased. But the work would not have been undertaken except for the need for improving the cold storage of fruit.

A similar illustration is afforded by the work of the insect department of the Natural History

Museum. Most people would say, I suppose, that the existence of men who spend their whole days examining and classifying and describing various small insects is, to put it mildly, highly academic. Yet in point of fact this is far from being the case. The growth of the great natural history museums in the nineteenth century had a very practical aim. It was based on the necessity for improving health and agriculture, especially in the tropics. The first necessity is to know what insect you are dealing with. It is no good knowing that malaria is transmitted by mosquitoes unless you know which one out of the several thousand existing kinds of mosquitoes transmits it, and what the habits of that particular one are.

The way in which the problem of the pest of the prickly pear in Australia was solved provides a good example of the work made possible on the basis of descriptive work by museum specialists. The plant was introduced into northeastern Australia in the last century, and rapidly spread until large areas of land became covered with forests of prickly pear. I remember a lecture by Dr. Tillyard, the Australian entomologist, in which he was talking about the prickly pear. In it he said, "I have been talking eleven minutes on the subject; in that time the prickly pear has occupied another eleven acres of Australian soil." It was a real menace. Then the authorities sent out an expedition to find out what insect enemies the prickly pear had in its natural habitat in

Central America. The expedition brought back a number of these; as a result of liberating them in vast numbers a great deal of the prickly pear forest has been done away with, and the pest is now under control. Without the previous work of museum specialists in systematic zoology, such results would be impossible.

Then there are people who hold that science is an evil genius, parent of unemployment. As an example of this aspect of science, they would quote modern machinery: for instance, an up-to-date automatic boot machine, one of the machines which enables a given number of men to turn out scores, and indeed hundreds, more boots than is possible by hand. The introduction of such machinery into manufacture must result in a certain amount of unemployment. But can one blame science for this? The scientist is asked to invent labour-saving machinery. He does what he is asked, and then he is blamed for saving labour! Surely the blame should be placed on the economic system which does not save labour in the form of increased production and increased leisure, but in that disagreeable form of enforced leisure known as unemployment.

War provides yet another illustration. Everyone has seen war photographs of the sufferings of victims of poison gas. It is true that poison gas in war would not exist but for science, but what are the precise facts? All the gases used on a large scale in the Great War were, as a matter of fact, dis-

covered in the ordinary routine of academic chemistry well back in the nineteenth century. It was the
military and political authorities, much later, who
suggested and who permitted their use in warfare,
not the scientists. This brings up a point of great
importance, namely, that science, like any other
tool, cannot be regarded as inherently good or
bad. Science is essentially an instrument of
control, a tool, and you cannot ascribe moral
values to a tool. You may use a hammer to nail
up a picture or to smash someone's skull, but it
is not the hammer which is good or bad, it is
your action and your motives. In the same way
science has been often used for good purposes,
and often for evil purposes, but it is not science
which is good or bad in either case.

Let us also remember that the use of scientific
work to increase the efficiency and the horrors
of war always has a set-off in the shape of some
results of peace-time application. For example, a
most efficient modern gas-mask is employed not
for military use, but for mine rescue work,
but it has been perfected as a result of the work
of the gas-defence research station at Porton.
However, such results are only a set-off. To
rely on the by-products of war research is
not an ideal method, of course, of getting
peace-time results. I do not recommend it any
more than I would recommend the method
employed by the Chinaman in the *Essays of Elia*
for obtaining roast pork. He found his house

burnt down with a little suckling pig in it. The roast pig he discovered was very delicious, and so he went about buying houses, putting pigs in them, and burning them down. We could get results of social value much more quickly and more efficiently by aiming at them directly.

Now we come to examples of misapplied or unapplied science. Take the subject of building; a great deal of valuable research is going on in this field. Among other things, this type of research involves tests for standardization. Tests have been made with concrete beams to find out their breaking-strain points. Only by such methods can you keep material up to standard.

Another example of standardization is carried out in what is known as the brick's cemetery. A number of samples of bricks are stuck in the earth and left there for a number of months; in this way the less-resistant sorts can be detected and rejected.

A great deal of other research is going on, aimed at increasing the efficiency and comfort of modern dwellings. Yet we know that that research is by no means fully applied. We still see in many of our great cities slum houses that can only be described as horrible. Yet when it comes to housing gorillas at the institution of which I have the honour to be secretary, we give them a beautiful building with every comfort and convenience. I am not upholding this on moral grounds, nor because I hold any special brief for gorillas. We do it because there is an obvious economic motive. We must keep our animals well-protected and well exhibited in attractive buildings in order to attract the public to the gardens.

Another illustration can be taken on the subject of the lack of application of scientific knowledge to health. It concerns the importance of the accessory food factors or vitamins. There were two twin baby sisters after the War in Vienna, both terribly undernourished. Room was found for one in a hospital, where she was properly fed, but there was no room for some months for the other. The contrast in weight and health was astonishing.

Recent research has shown that even in the case of healthy boys that you would think well above the average, the addition of extra milk will promote extra growth and vigour. We are very far from having reached the optimum of physique. It is perfectly true that the average has been raised. However, we are told that only a certain small proportion of the country's children are below standard; this is quite misleading. The word "standard" is here being used in an unscientific sense to mean merely the existing average. But if the average is so far below the optimum it can hardly be called a standard.

Investigations on the pollution of the atmosphere by smoke have produced some instructive information. Below I set out a curve which relates to Manchester. You see that deaths from respiratory diseases go up and down with the number of days of fog per week. During the

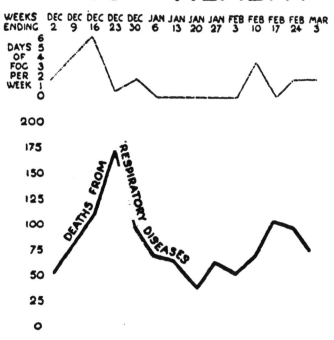

The above diagram (prepared in the Manchester Public Health Department) illustrates a period of fog in Manchester. The thin line represents the number of days of fog per week, rising at the highest point to six days out of seven. The thick line indicates the deaths from respiratory diseases, pneumonia &c, immediately following the same period.

FIG. I.

investigation photographs were taken on two different days of the atmosphere of a certain industrial city. The photographs taken on the

Sunday showed that the air was nice and clear. Those taken on Monday morning, when the industries were starting up their week's work, shows the pollution and fog commencing. But the photographs taken on the Monday afternoon showed that the pollution and fog had become so dense that objects visible in the earlier pictures were hardly discernible. Really, a visitor from another planet would be pardoned for thinking that we were not sane, if he came down to earth in some inter-planetary ether-ship and saw these palls of fog and smoke-cloud over all the major conglomerations of population in this country.

In this matter of smoke, again, science is *not* being applied. But there are examples of applied scientific method. A full-scale model of the parts of Piccadilly Underground Station was erected by the London Passenger Transport Board to find the optimum flow of passengers. Every time we go by Tube we are profiting by this application of scientific method. This is possible with big concerns under unified control. But when too many agencies are involved, then since we still have no real brain or central co-ordinating organization for research, you do not get scientific method.

We may take as an instance one of the great arterial roads out of London, on which millions of pounds have been expended. Owing to lack of co-operation between different authorities, ribbon building was allowed to go on until now, as every motorist who has motored along it is painfully aware, it will shortly be necessary to "by-pass the by-pass."

Research with regard to the problem of noise illustrates again the comparatively poor stimulus to research or its application provided by the consumer as against the producer. As all we city dwellers experience in our persons, noise is becoming one of the major problems of our civilization. It is true that a great deal of research is going on on the subject of noise—research on noiseless motors in industrial concerns; research by the London Passenger Transport Board in regard to noise in tube trains; by the Air Ministry on noise in aeroplanes; and so on and so forth. But there is no central Noise Research Station for the whole country, whereas every major industry is encouraged by the Department of Scientific and Industrial Research, and has its central laboratories, where all problems of the industry are considered in conjunction. But the consumers of noise—those who hear it—do not get any profit out of the noise-reducing instrument or machinery. Thus, until we, as consumers, can organize and bring pressure to bear we shall not get that centralized research which is so desirable.

Agriculture is the most obvious example of frustrated science. For example, at the Empire Grassland Research Station at Aberystwyth all kinds of grass are grown under controlled conditions, and experimental breeding is being carried out. Here Professor Stapledon has perfected a remarkable method, already demonstrated on quite a large scale, by which any area of mountain moorland could be converted into nice pasture with clover and good grass. You plough the ground up, fertilize it with the right mineral fertilizer, plant it with grasses that he has specially bred, and put sheep on it. By this method it would be possible to add good pasture to the tune of something like 15 per cent of the total area of these islands—a rather large application of science, in itself capable of doubling the number of sheep in the country. But I do not suppose there is the remotest chance of it being done on a large scale, because under the existing economic system the bottom would drop out of the market in sheep.

I referred earlier to population; let us just see why that subject is so important. The graphs here set out are of the future population trend of Britain. You can by comparatively simple calculations prophesy with reasonable accuracy what will happen to our population for about a generation ahead. You see our population will reach a maximum in a few years' time, and then go down. More important is the fact that the proportion of the various age groups will change. For instance, the proportion of young people under twenty will go down from about one-third to one-quarter of the total, while the proportion of old people above fifty-five will go up from one-sixth to about one-quarter. The whole basis

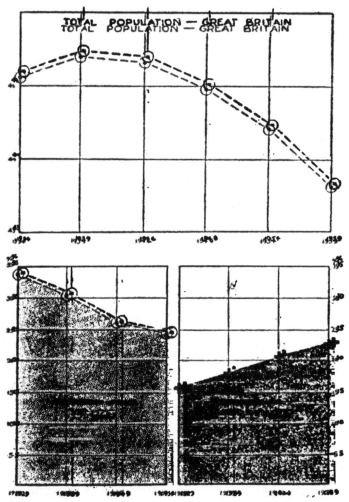

Above, a curve showing the probable total population of Great Britain, assuming no net emigration, for the next 25 years.

Below, the probable change in the proportion of young people up to the age of 20 (left) and of old people above the age of 55 (right), calculated as percentages of the total population. (Based on figures given in *Planning*, No. 4, June 6, 1933.

FIG. 2.

of social and family life will be different. This concerns every branch of national life:—education, the number of children entering industry each year, the size of classes, old-age pensions, the military services, the labour force—almost everything you can think of. You would suppose that research on that subject would be intensively fostered by the Government, and yet, as I say,

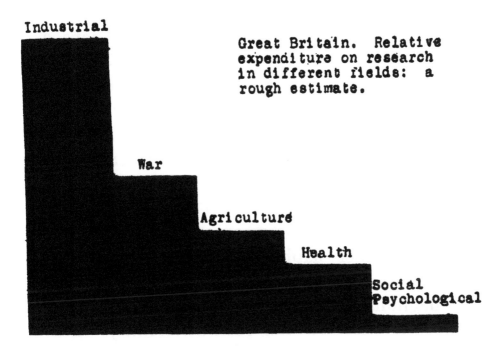

Great Britain. Relative expenditure on research in different fields: a rough estimate.

FIG. 3.

it has remained for private academic scientists to work out and publish such important facts.

Now, finally, in order to emphasize what I have said of the lop-sided nature of our research structure I append a figure. This picture gives you roughly the proportionate amounts, so far as they can be estimated, of money spent on different branches of scientific research. As I pointed out, this is just the reverse of what one would imagine was theoretically desirable.

The National Physical Laboratory at Tedding-
ton provides an example of the scale on which
physical research is done in this country. It is
a veritable city and harbours many more research
workers than all those in psychology and social
science in the whole country.

What can one say in conclusion? Perhaps a
brief glance at the history of science will not be
out of place. You start on ancient science, with
its somewhat haphazard advance; then you have
the blank of the Dark and Middle Ages. Science
first becomes organized in the seventeenth century, with a basis of deliberate experiment, of full
publication of methods as well as results, and of
learned societies. This was accompanied by the
rapid, if rather haphazard, application of the
results of pure science. Quite recently, say from
the beginning of this century, a further phase
has begun. Research has been deliberately organized. People have realized that practical problems
can be solved by handing them over to the pure
scientist, even if for a time his work seems to
have no relevance to practice, and basic research
has been interposed as a link in the chain between
question and answer. We may speak of this
phase in the history of science as the phase of
research institutions, some of them private, some
of them academic, and some of them under the
State. So far this method has been applied in
industry, war, agriculture, and medicine.

Perhaps we may date a further phase from the

beginning of the world crisis. Since then, many people have been thinking very fundamentally about science, and have come to the conclusion that it is frustrated and lop-sided, and that it is so because it is a function of an inadequate economic and social system. They are beginning to realize that scientific research is a link in the chain between question and answer in the great social problems which press upon us. We need more research on human nature and on social structure. Let us not forget that it is quite untrue to say that human nature cannot change. It is always changing. It changes according to the system of ideas and the economic and class system in which we are brought up; we should all of us be quite different if we had been brought up under a different régime.

Do not let us forget that average men and women have in their nature all sorts of potentialities of which we hardly dream to-day. We cannot prophesy the possibilities of artistic appreciation and execution, of intellectual achievement, perhaps of the so-called supernormal faculties. Research is needed to determine the degree of those possibilities: but I would be pretty sure that they are much greater than most people now imagine.

Research would discover enormous new potentialities, not only of individual human nature, but also of group behaviour. For that we want research on pyschology, and in particular a truly scientific study of education instead of the approach, based on tradition, religious motives, and State needs, which is the usual approach to-day. We also need research on such objects as population and its regulation, and the improvement of the human stock. There is also research into the causes of war. Let me remind you that millions of pounds are spent on research aimed at improving the efficiency of war, but so far as I know no money is spent on research on how to prevent war.

Social science demands new methods of study, but I have no time to go into that. In any case we shall not develop them unless we try. Then, too, in the application of social science there will be a fight against the laziness of human nature, vested interests, greed, organized group sentiment, superstition, and so on; but that is no reason for not undertaking the fight.

The picture which I have drawn may perhaps seem a little gloomy, of science largely enslaved by the profit-motive and distorted by various pressures in the social system, but all the same I do myself firmly believe that it is the only tool on which we can rely to get us out of our difficulties. Of course, it is only a tool. Besides science you need goodwill, you need heart as well as head; but you need the right tool, and science has proved itself to be the only tool for getting long-range results of a fundamental nature in the later stages of man's struggle for control over Nature. We must go forward. I said at the beginning that this age was usually spoken of as a scientific age. That is not true. It is merely an age with a scientific basis. We need scientific superstructure too. You cannot mix two attitudes of life, one scientific concerning external nature and another unscientific or pre-scientific concerning human nature, without disastrous results. It is more serious to mix your theories of life than to mix your drinks.

For the scientist the next step is something he has as yet hardly envisaged. It is the scientific study of science itself, as a social function—just as a physiologist might make a scientific study of digestive or nervous function. And this study will, in the long run, allow the application of science to be controlled and developed along the best possible lines.

In any case, whatever our present troubles, it is largely through science that we have reached a level where we can look forward hopefully. Science has opened all kinds of perspectives. In particular it has shown life as a slow, upward-evolving process. It has shown that there is something in evolution which we must call progress and it has shown that we ourselves are now trustees for any evolutionary progress that remains to be made. I think the biologists can definitely say that if there is to be a continuation of the upward progress of evolution it will only be by means of man's conscious control of the process. He has got to take over from the blind,

wasteful forces of non-human nature. Already
in the brief space of human history we see man
creating truth. Life in man is bringing values to
birth. History does not just happen: we make it
with our thoughts and ideas, and with our
material progress.

We have—some people regret it—left definitely
behind us the old age of faith inspired by belief
in authority and revelation, but through the
perspectives opened by science I think it is
possible to look forward to a new age of faith—
faith that by effort and will which is linked to
right ideals and based on real knowledge we
can create newer and better worlds for life. And
I think the scientist would emphasize that
without more science and better science we
cannot hope for that progress. We could control
human nature, we could leave a better world for
our grandchildren; but we can only do this if we
improve scientific methods and their application.
That, I think, is the long-range view to take of
science in its relation to social needs.

9 781138 981461